# ADVANCES IN MULTI-PHOTON PROCESSES AND SPECTROSCOPY

# ADVANCES IN MULTI-PHOTON PROCESSES AND SPECTROSCOPY

Volume 18

Edited by

**S. H. Lin**
*National Chiao-Tung University, TAIWAN*
*Institute of Atomic and Molecular Sciences, TAIWAN*
*and Arizona State University, USA*

**A. A. Villaeys**
*Institute de Physique et Chimie des*
*Matériaux de Strasbourg, FRANCE*

**Y. Fujimura**
*Tohoku University, JAPAN*

NEW JERSEY • LONDON • SINGAPORE • BEIJING • SHANGHAI • HONG KONG • TAIPEI • CHENNAI

*Published by*

World Scientific Publishing Co. Pte. Ltd.
5 Toh Tuck Link, Singapore 596224
*USA office:* 27 Warren Street, Suite 401-402, Hackensack, NJ 07601
*UK office:* 57 Shelton Street, Covent Garden, London WC2H 9HE

**British Library Cataloguing-in-Publication Data**
A catalogue record for this book is available from the British Library.

**Advances in Multi-Photon Processes and Spectroscopy — Vol. 18**
**ADVANCES IN MULTI-PHOTON PROCESSES AND SPECTROSCOPY**



ISBN-13 978-981-279-173-3
ISBN-10 981-279-173-6

Typeset by Stallion Press
Email: enquiries@stallionpress.com

Printed in Singapore.

# Preface

In view of the rapid growth in both experimental and theoretical studies of multiphoton processes and multiphoton spectroscopy of atoms, ions, and molecules in chemistry, physics, biology, materials science, etc., it is desirable to publish an advanced series that contains review papers readable not only by active researchers in these area, but also by those who are not experts in the field but who intend to enter the field. The present series attempts to serve this purpose. Each chapter is written in a self-contained manner by the experts in the area so that the readers can grasps the knowledge in the area without too much preparation.

The topics covered in this volume are "Nonlinear optics for characterizing XUV/soft X-ray high-order harmonic fields in attosecond regime," "Signatures of molecular structure and dynamics in high-order harmonic generation," "Molecular manipulation techniques and their applications," "Sum frequency generation: an introduction with recent developments and current issues," "Propagation and intramolecular coupling effects in the four-wave mixing spectroscopy," and "Control of molecular chirality by lasers." The editors wish to thank the authors for their important contributions. It is hoped that the collection of topics in this volume will be useful not only to active researchers but also to other scientists in biology, chemistry, physics, and materials science.

S. H. Lin,
A. A. Villaeys,
Y. Fujimura

# Contents

Chapter 1

# Nonlinear Optics for Characterizing XUV/Soft X-ray High-order Harmonic Fields in Attosecond Regime

Yasuo Nabekawa* and Katsumi Midorikawa†

*Laser Technology Laboratory, RIKEN,*
*2-1 Hirosawa, Wako-shi, Saitama, 351-0198, Japan*
**nabekawa@riken.jp*
*†kmidori@riken.jp*

## 1. Introduction

### 1.1. *Nonlinear phenomena in XUV/soft X-ray region for ultrafast optics*

The nonlinear transition of electrons induced with an intense light field in the extreme ultraviolet (XUV) or soft X-ray region is an attractive phenomenon because it can be expected to reveal new features of the interaction between electrons and photons, due to the energies of the photons being sufficiently high to singly ionize an atom with one photon. From the viewpoint of the attosecond science[19] in which the duration of the optical pulses should be much shorter than 1 femtosecond, this phenomena is technically important to measure or to determine the characteristic of a so-called attosecond pulse, the wavelength of which necessarily comes in the region of XUV/soft X-ray due to the requirement of the shorter optical period than the pulse duration. However, experimental research into the nonlinear interaction of the ultrafast XUV/soft X-ray light pulses has been slow to progress, except for some pioneering works,[29,40,70] because of the lack of intense ultrafast light sources in this wavelength region.

A free electron laser (FEL) is a promising apparatus for generating intense, coherent, short-wavelength light. Wabnitz *et al.* demonstrated that the cluster of xenon atoms could explode with absorption of the XUV pulse generated from the FEL in DESY facility.[71] They also found that the yield of nitrogen ion was quadratically depend on the intensity of a XUV pulse from the FEL.[62] In spite of these fruitful outcomes of the FEL light source, the detail of the XUV pulse itself from the FEL is still unclear because it has not been directly measured.

The most favorable alternative to the FEL is the high-order harmonic field generated with femtosecond laser pulses.[11,34,37,38,56] The pulse energy of the high-order harmonic field is usually lower than that of the FEL pulse, while the shorter pulse duration of the high-order harmonic field, which can be shortened to attosecond regime, can compensate the lower energy to obtain higher intensity than that of the FEL pulse. The first demonstration of the nonlinear interaction of the high-order harmonics with an atom was the side-band generation in photoelectron spectra from a rare gas atom dressed with an intense visible laser field, which is the so-called two-color above-threshold ionization (two-color ATI).[17,57] Application of the two-color ATI for characterization of the pulse of the harmonics,[5,9,17,58] has been feasible. The side-band generation, however, is mostly owes to the high intensity of the visible laser pulse. The intensities of the high-order harmonics have been too low to induce the nonlinear effects, so that the two-color ATI remains a process with a single photon in the harmonics.

Thus, the nonlinear transitions induced with the high-order harmonic field have themselves attracted growing interest in physics. In addition, they are essential to directly observe the characteristic of an optical pulse. In the next subsection, we will briefly review the conventional autocorrelation technique for characterizing a femtosecond laser pulse in the visible region in order to clarify the importance of the nonlinear interaction, and introduce a short history of the pulse measurement of high-order harmonic fields.

## 1.2. *Autocorrelation measurement*

In the autocorrelation measurement, the measured optical pulse should be sent to an interferometer, in which the pulse is split into two replicas with a half mirror. The two replicas of the pulse are combined on another half mirror in the interferometer, then the combined pulse is focused into a nonlinear medium, as shown in Fig. 1. We can find the correlated signal as a function of delay between the two replicas by observing the outcome of the nonlinear interaction, such as the second harmonic pulse, the self-diffracted pulse with a transient grating, the multi-photon fluorescence, the ion yield with multi-photon absorption, and

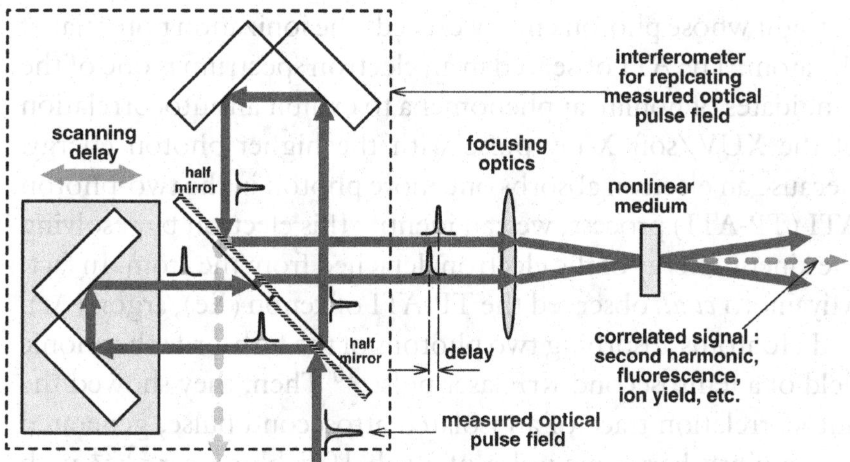


**Fig. 1.** Schematic figure of autocorrelation measurement.

so on. Any linear interactions cannot provide the information of the temporal profile for the energy flow of the optical pulse. They only indicate how long the optical field is coherent in time domain.

Hence, the search for the nonlinear phenomena is inevitable for researchers who want to know directly the temporal profile of the high-order harmonic fields. Two-photon absorption (TPA) in an atom of a gas medium might be the most favorable nonlinear phenomenon because we can detect the ions or the electrons yielded via TPA by conventional time-of-flight spectroscopy. The first observation of TPA induced by the XUV harmonic field of a femtosecond Ti:sapphire laser was reported by Kobayashi *et al.*[29] They revealed that the yield of a singly charged helium ion ($He^+$) should be due to the TPA of the ninth-order harmonic field, and then demonstrated the autocorrelation measurement by observing the $He^+$ yield, resulting in the duration of the ninth-order harmonic field being 27 fs. It was another five years before Tzallas *et al.* found the autocorrelation trace of multiple harmonic fields in a similar manner.[70]

The detection of singly charged ions of rare gas atoms, in principle, is not suitable for demonstrating TPA of XUV/soft X-ray light whose photon energy exceeds the ionization potential of the atom. The ATI observed in an electron spectrum is one of the candidates of nonlinear phenomena to exhibit an autocorrelation of the XUV/soft X-ray pulse with the higher photon energy. Because an electron absorbs one more photon in the two-photon ATI (TP-ATI) process, we can identify this electron by resolving the kinetic energy of the electron detached from the atom. In fact, Miyamoto *et al.* observed the TP-ATI of xenon (Xe), argon (Ar), and He atoms absorbing two photons of the fifth-order harmonic field of a femtosecond KrF laser pulse.[40] Then, they showed the autocorrelation trace of an isolated attosecond pulse, generated as the ninth-harmonic pulse of a sub-10 fs blue laser,[23,24] with the TP-ATI of He.[31,60]

Our report on the autocorrelation measurements of the high-order harmonic fields in this chapter are based on the similar observations of the nonlinear interactions, while we should note that the successful results of our experiments owe to the highest intensity of each harmonic field, to our knowledge, obtained from our distinctive scheme of the harmonic generation. We will briefly present how to generate the intense harmonic fields before describing their application to the nonlinear optics for the autocorrelation measurement.

## 2. Generation of Intense Harmonic Fields

The schematic view of the experiment for the high-order harmonic generation of a femtoseocnd visible–near infrared laser pulse is depicted in Fig. 2(a).

The femtosecond laser pulse is focused into a rare gas target supplied with a pulsed gas valve synchronously triggered with the laser pulse, or with a static gas cell having an entrance-pinhole and an exit-pinhole. The typical intensity of the laser pulse is approximately $10^{14}$ W/cm$^2$, although it depends on the gas species and the harmonic orders for the maximum yields. The generated high-order harmonic fields co-propagate with the pump laser pulse in a vacuum ambient condition in order to eliminate the absorption by the air.

Typical harmonic spectra generated from a neon gas target is shown in Fig. 2(b).[66,67] We can see the similar intensity of each harmonic spectrum from the 33rd order to the 63rd order, which is the specific feature of the high-order harmonic generation. The range of the harmonic orders having such similar intensities is called "plateau" region, while the harmonic order at which the intensity rapidly decreases is named "cut off."

We have to notice how the harmonic fields emerge from the rare gas target to realize the high-intensity light source of

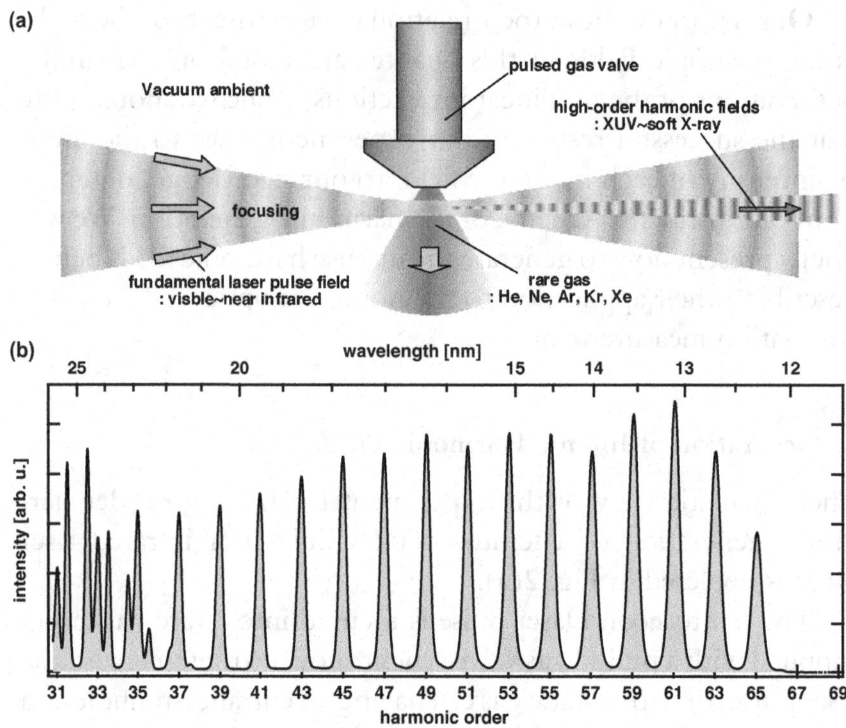


**Fig. 2.** (a) Schematic view of high-order harmonic generation. (b) Typical spectra of the high-order harmonic fields generated with neon gas.

those fields. The physical process of the harmonic generation is classified into two categories; the first of which is the single atom response of the rare gas in the intense laser field. The second is a macroscopic effect during the propagation of the harmonic fields in the rare gas as a dispersive and absorptive medium.

### 2.1. *Single atom response*

In the quantum description of an electron in a rare gas atom interacting with a classical linear polarized electric field under the dipole approximation, Lewenstein *et al.* simplified the issue for an induced electric dipole moment, which is a source of the harmonic fields, originating from the electron motion.[33] They neglected all the intermediate states of the electron and considered only the bounded ground state without the electric field and the continuum state under the electric field (Strong Field Approximation, SFA). Under the SFA condition, the time-dependent Shrödinger equation can be analytically integrated because the dependence of the spatial coordinates of the electric field can be excluded owing to the fact that the wavelength of the visible–near infrared light field is much longer than the typical scale of a rare gas atom.

Thus, one can find the Fourier transform of the induced electric dipole moment should be proportional to the product of: (i) the transition dipole matrix element from the ground state to the continuum state at time $t'$, (ii) the applied laser field, (iii) the transition dipole matrix element from the continuum state to the ground state at time $t$, and (iv) the phase factor corresponding to the action that describes the classical motion of the electron in the electric field without the effect of the bound potential of the atom.

We can correlate these terms with the classical motion of the electron in the so-called three-step model for the high-order harmonic generation proposed by Corkum.[8] First, the electron should be departed from the atom (ionization of the atom) corresponding to the terms of (i) and (ii), then, the electron goes on excursion in the laser field with receiving energy as the phase factor expressed by the term (iv), and finally, it should go back to the ion (recombination of which amplitude is equal to the term (iii)) with the emission of the harmonic fields. These three processes are schematically shown in Fig. 3. We can find, by applying the three-step model, that the photon energy at the cutoff of the harmonic field should agree with the summation of ionization energy and the maximum kinetic energy of the electron at the moment of the recombination. Thus, this conclusion of the

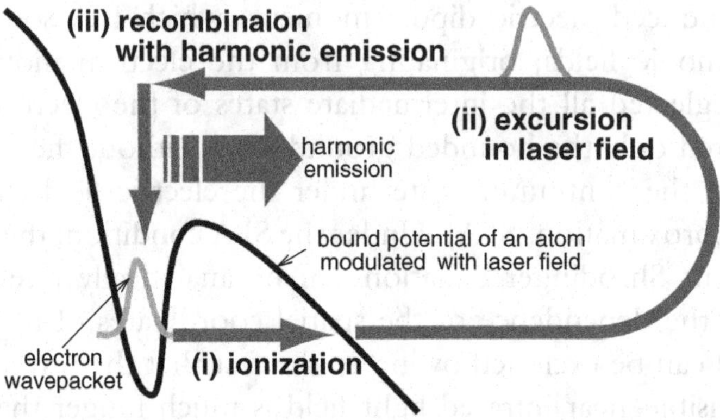


**Fig. 3.** The three-step (classical) model of high-order harmonic generation. These process can be related with the quantum description of electron motion under the SFA condition.

classical model should also be justified by the quantum model with the SFA.

In fact, the four terms in the induced dipole moment are 5-fold integrated with respect to $t$, $t'$, and the three-dimensional momentum of the continuum state reflecting the accumulation of the all possible quantum paths, while the significant contribution to those integrations comes from the region at which the phase, namely the action does not rapidly change for those variables. Hence, we can easily execute the integrations by substituting the action to that of the second order approximation in Taylor series at the stationary points, at which the first derivatives of the action to the variables should be equal to zero. The integrals result in the contour integrals on the complex plane, so that the stationary points are generally complex. From this stationary point analysis,[33] we can see the important result that the spectral phase of the harmonic components in the induced dipole moment should be $-\alpha I - \hbar^{-1} I_p \tau_e$, where $I$ is the intensity of the laser field, $I_p$ is the ionization energy of the atom, $\tau_e$ is the excursion time of the electron defined as time difference between $t$ and $t'$ at the stationary points, and $\alpha$, which is called reciprocal intensity, is a constant depending on $\tau_e$. Thus, the intensity of the laser field and the excursion time significantly affect the characteristics of the harmonic fields.

In particular, the excursion time is not uniquely determined for the harmonic fields in the plateau region due to the periodicity of trigonometric functions on the complex plane. As a result, two kinds of $\tau_e$'s, originating from the "short" and the "long" trajectories during the excursion, respectively, mainly contribute the phase,[3] and therefore the temporal profile of the synthesized harmonic field in the plateau region is not expected to be an elementary pulse train such as the laser field generated from the mode-locked oscillator.

On the other hand, the two $\tau_e$'s merge into the unique at the harmonic order higher than the cutoff, so that the temporal profile can be clearly defined by selecting the wavelength range of the cutoff with a filter and/or a multi-layered mirror as was demonstrated in Refs. 19, 26, and 55.

We have known the principal natures of the induced dipole moment of the electron in the single atom as a point emitter of harmonic fields. The observed harmonic fields in an experiment, however, is always the consequence of the coherent sum of each harmonic fields coming from each emitter pumped by the laser field, which propagates with the yielded harmonic fields. The key issue for generating intense harmonic fields relies on the successful accumulation of the emission at each harmonic source with a macroscopic effect.

### 2.2. *Propagation of the harmonic fields with pumping laser field: Phase matching*

In the macroscopic point of view, the nonlinear polarization in a rare gas medium at $(2n+1)$th order harmonic field at a position of $\vec{r}'$, notated as $\widetilde{P}_{\mathrm{NL}}^{2n+1}(\vec{r}')$, should be equal to the product of the positive frequency part of the Fourier component of the induced dipole moment at the $(2n+1)$th order, $\widetilde{\mu}^{+}_{2n+1}(\vec{r}')$, the density of atoms, $\rho$, and the phase factor accompanying the propagation of the pumping fundamental laser field, $e^{i(2n+1)\varphi_f(\vec{r}')}$, namely,

$$\widetilde{P}^{2n+1}_{\rm NL}(\vec{r}') = \rho\widetilde{\mu}^{+}_{2n+1}(\vec{r}')e^{i(2n+1)\varphi_f(\vec{r}')}, \qquad (1)$$

where $\varphi_f(\vec{r}')$ is the phase of fundamental laser field.[35] We adopt the scalar approximation for the nonlinear polarization and the electric field of the harmonic because of the linear polarization. The electric field of the $(2n+1)$th order harmonic is evolved in accordance with the Helmholtz equation in the frequency domain with the source term of $-\varepsilon_0^{-1}\vec{k}^2_{2n+1}\,\widetilde{P}^{2n+1}_{\rm NL}(\vec{r}')$. We denote the dielectric constant of vacuum as $\varepsilon_0$ and the wave vector of the $(2n+1)$th harmonic field as $\vec{k}_{2n+1}$. Thus, the observed $(2n+1)$th order harmonic field at a position of $\vec{r}$, $\widetilde{E}^{2n+1}(\vec{r})$, can be written as

$$\widetilde{E}^{2n+1}(\vec{r}) = -\varepsilon_0^{-1}\vec{k}^2_{2n+1}\int_V d^3\vec{r}'\,G_R(\vec{r}-\vec{r}')\widetilde{P}^{2n+1}_{\rm NL}(\vec{r}'). \qquad (2)$$

The well known retarded Green function of the Helmholtz equation, noted as $G_R(\vec{r}-\vec{r}')$ in this equation, includes the phase factor of the outgoing spherical wave, $k_{2n+1}|\vec{r}-\vec{r}'|$, while the nonlinear polarization gives the integration of the phase factors of $-\alpha I(\vec{r}')$ coming from $\widetilde{\mu}^{+}_{2n+1}(\vec{r}')$ and $(2n+1)\varphi_f(\vec{r}')$ arising from the propagation of the fundamental laser field, respectively. We added the dependence of the macroscopic coordinates of $\vec{r}'$ to $I$ in $\widetilde{\mu}^{+}_{2n+1}(\vec{r}')$.[4] The main contribution of this volume integration in the effective volume of the gas medium, $V$, should originate from the stationary contour for the total phase of the above-mentioned three, and therefore, we can obtain the generalized "phase" matching condition by setting the differentiation of the total phase with respect to $\vec{r}'$ to zero, although the dimension of the equation does not correspond to that of the "phase".

$$\vec{0} = -k_{2n+1}\widehat{e}(\vec{r}') + (2n+1)\vec{k}_f(\vec{r}') + \vec{K}_d(\vec{r}'), \qquad (3)$$

where we define $\vec{k}_f(\vec{r}') \equiv \vec{\nabla}'\varphi_f(\vec{r}')$ and $\vec{K}_d(\vec{r}') \equiv \vec{\nabla}'\{-\alpha I(\vec{r}')\}$. The unit vector $\vec{e}(\vec{r}')$, of which direction is determined by minimizing the magnitude of the wave vector difference (phase mismatch) to the variable $\vec{r}$ coordinate, results in being parallel to $(2n+1)\vec{k}_f(\vec{r}') + \vec{K}_d(\vec{r}')$.[4] Equation (3) is similar to the conventional phase matching condition for the harmonic generation except for the dipole phase in the single atom response, $\vec{K}_d(\vec{r}')$, which plays an important role to determine the condition for the efficient harmonic yield.

Hereafter, we assume the focused Gaussian beam profile of the fundamental laser field. Thus, the magnitude of the local wave vector of $(2n+1)\vec{k}_f(\vec{r}')$ should be smaller than $k_{2n+1}$ because the correction of the wave vector coming from the geometrical phase in the focused fundamental laser field is always directed to the opposite to the propagation direction and it can overcompensate the effect of larger refractive index for the fundamental laser field than that for the high-order harmonic fields in a gas medium. With this restriction of the wave vectors, the phase (wave vector) matching condition can be satisfied at the region where the dipole wave vector is nearly parallel to $\vec{k}_f(\vec{r}')$, as shown in Fig. 4(a), if the magnitude of the dipole phase is suitably small.[4] We can confirm, from the stationary point analysis with an appropriate intensity of the laser field, that the dipole phase originating from

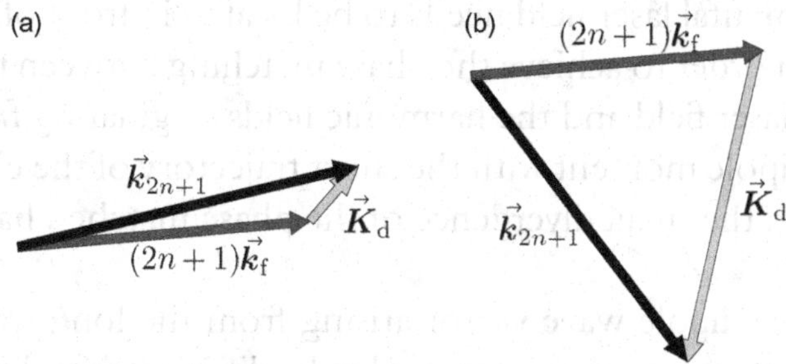


**Fig. 4.** Phase (wave vector) matching condition for (a) the small dipole phase (wave vector) originating from the short trajectory of the electron and for (b) the large dipole phase (wave vector) originating from the long trajectory of the electron.

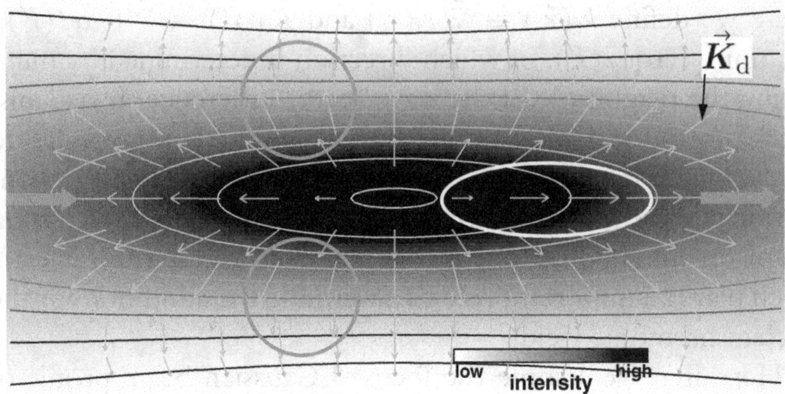


**Fig. 5.** Schematic figure of the intensity of the laser field near the focus (image with a gray scale) and the dipole wave vector at each point (arrows). A thick gray arrow indicates the propagation direction. The region suitable for the phase matching conditions are surrounded with a solid white ellipsoidal for the short trajectory, and with gray circles for the long trajectory, respectively.

the short trajectory of the electron is well situated upon the relation among these three wave vectors. Because the direction of the dipole wave vector points the way of the decreasing intensity due to the minus sign in front of the dipole phase, the region satisfying the phase matching condition is limited behind the focus of the fundamental laser field. The intensity and the direction of the dipole wave vectors in the vicinity of the focal region is schematically shown in Fig. 5. In other words, the focal point of the fundamental laser field needs to be located in front of the gas medium in order to achieve the phase matching between the fundamental laser field and the harmonic fields originating from the induced dipole moment with the short trajectory of the electron, resulting in the small divergence of the phase matched harmonic fields.[4]

Another dipole wave vector arising from the long trajectory of the electron gives us a completely different condition for the phase matching. Because the reciprocal intensity $\alpha$ for the long trajectory is much larger than that for the short trajectory (typically, one order of the magnitude larger), the magnitude of $\vec{K}_{\mathrm{d}}(\vec{r}')$ is notably large. The direction of $\vec{K}_{\mathrm{d}}(\vec{r}')$ should be almost perpendicular to the propagation axis and the wave vector of the harmonic field should point far from $\vec{k}_{\mathrm{f}}(\vec{r}')$, as shown in Fig. 4(b). The region on which $\vec{K}_{\mathrm{d}}(\vec{r}')$'s having such characteristic are lying is located behind the focus and far from the propagation axis of the fundamental laser field, as indicated with gray circles in Fig. 5. Therefore, we can observe the phase-matched harmonic fields from the long trajectory of the electron, by adjusting the focal position behind the gas medium, with a large divergence having a spatial profile of a torus.[4]

One of the most important consequences of the above-mentioned analysis of the phase matching is that the mixture of the two phase properties in the harmonic fields induced in a single atom response can be filtered with the macroscopic effect of the phase matching during the propagation. One can skim the short-trajectory harmonic emission from the dipole moment by adjusting appropriately the focal position in front of the gas medium. The spatial profile of the resultant harmonic beam is Gaussian-like with a small divergence and the spectral phase is well-defined. The long-trajectory harmonic emission is weak due to quantum diffusion of the trajectory during the long excursion time. Further, the ring-like shape of the spatial profile does not suitable for applications. Hence, we should concentrate our interest on the phase-matched harmonic fields of the short trajectory in the following sections.

### 2.3. *Development of intense high-order harmonic generator*

How can we efficiently generate the high-order harmonic fields in practice?: (i) by increasing the intensity of a pumping fundamental laser field? or (ii) by increasing the density of a target gas medium? or (iii) by stretching the medium length of the gas target?

Higher intensity of the laser field than the complete ionization threshold of the gas target depletes neutral atoms as emitters of the high-order harmonic fields and disturbs the propagation of the laser field due to the additional refractive index gradient induced with plasma, resulting in decreasing the harmonic efficiency. Thus, we should restrict the intensity within a range of ionization limit, that is typically $10^{14}\,\mathrm{W/cm^2}$ to a few $10^{14}\,\mathrm{W/cm^2}$ depending on the atom species.

It is somewhat complicated to conclude whether the last two candidates of the answers are true or not. In fact, the more gas density, or equivalently the more pressure, the more atoms can be interacted with the laser field, while the change of the gas density significantly affects the phase matching condition and the absorption. The longer medium length also increases the number of the interacting atoms, but causes the higher absorption.

We will simplify the phase-matching integral in Eq. (2), according to the literature written by Constant *et al.*,[7] by assuming the laser field to be loosely focused in the gas medium such that the transverse component of the wave vectors can be neglected and the intensity of the laser field does not notably change, neither does the dipole $\widetilde{\mu}^+$ change, in the target gas medium, in order to see the substantial effect for these parameters. Under this approximated condition, the integral with respect to the transverse components can be easily carried out, yielding the factor including the cross section of the laser field, $\pi w^2$, where $w$ is the $e^{-2}$ radius of the beam. The remaining integral for the propagation axis, $z$, is expressed as $F \equiv \int_0^L dz e^{(-\gamma+i\Delta k)z}$, so that the harmonic yield, $I_\mathrm{h}$, should be written as

$$I_\mathrm{h} \propto \left|\widetilde{\mu}^+\right|^2 \rho^2 w^2 \left|F\right|^2 . \tag{4}$$

We define the notations of the medium length, the phase (wave number) mismatch, as $L$ and $\Delta k$, respectively. We introduce the absorption coefficient, $\gamma$, as a imaginary part of the wave number

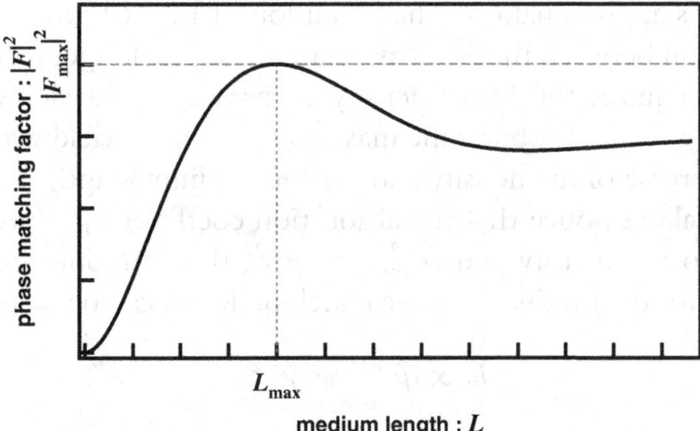


**Fig. 6.** Phase matching factor $|F|^2$ for the loosely focused laser field as a function of medium length. The appropriate medium length for the maximum $L_{max}$ should be near the half of coherence length $\pi/\Delta k$.

of the harmonic field by hand. The phase-matching factor of $|F|^2$ is obtained as a function to the medium length.

$$|F|^2 = \frac{1}{\gamma^2 + \Delta k^2}\{1 + e^{-2\gamma L} - 2e^{-\gamma L}\cos(\Delta k L)\}. \qquad (5)$$

Thus, this factor reaches its maximum $|F|^2_{max}$, which is approximately equal to $\gamma^{-2}$, at the medium length being approximately equal to $L_{max} = \pi/(2\Delta k)$ under the condition of $\Delta k \ll \gamma$, as shown in Fig. 6. Because the most part of the phase mismatch (neglecting the geometrical phase) originates from the mismatch of the refractive index of the harmonic field to that of the fundamental field due to the dispersion of the gas medium, $\Delta k$ is approximately decomposed into the product of the density $\rho$ and the phase mismatch per unit density, defined as $\overline{\Delta k}$, which is a constant for gas species. We can find the simple relation of the density and the medium length for the maximum phase matching factor, namely,

$$\rho_{max} L_{max} \sim \pi/(2\overline{\Delta k}) = \text{const.}, \qquad (6)$$

from this approximation. The condition of Eq. (6) gives us the restriction between the two gas parameters. The longer medium length requires the lower density or pressure, and vice versa. Hence, we cannot obtain the maximum harmonic yield with the over-increase of the density and/or the medium length.

By taking notice that the absorption coefficient $\gamma$ is proportional to the density and $|F|^2_{\max} \sim \gamma^{-2}$, the harmonic yield in Eq. (4) for the maximum phase matching factor can be written as

$$I_{\mathrm{h}} \propto \left|\widetilde{\mu}^{+}\right|^{2} w^{2}\overline{\gamma}^{-2}, \tag{7}$$

where $\overline{\gamma}$ is the absorption coefficient per unit density, which is fixed to the gas species.

We can find in Eq. (7), at a glance, that the density is canceled out. Thus, *only we can do for the increase of the harmonic yield is to broaden the cross section, $w^2$, of the fundamental laser field* with an appropriate intensity. This conclusion is very simple but clearly indicates the way we should go.

Based on the above-mentioned analysis, we have developed the intense high-order harmonic generator. Schematic figure of the setup is depicted in Fig. 7(a).[64,67] The pumping fundamental laser pulse is delivered from a chirped pulse amplification (CPA) system of Ti:sapphire laser with a pulse energy of 10–50 mJ,

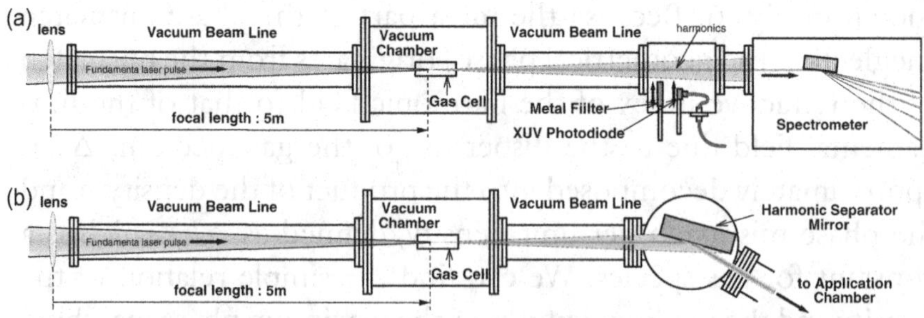


**Fig. 7.** Schematic figure of the setup of the intense high-order harmonic generator. The most distinct feature of this system is the long focal length of the laser field. (a) Configuration for the measurement of the harmonic spectra and pulse energy, and (b) that for the applications.

a pulse duration of 30–40 fs, and a repetition rate of 10 Hz. The laser pulse is focused by using a positive lens with a focal length of 5 m, which is several times longer than the typical focal length for the harmonic generation. When we need to eliminate unfavorable nonlinear effects, such as a self-focusing or a self-phase modulation, caused by the air and/or the glass material of the lens, the grating compressor in the CPA system and all the focusing optics should be located in a vacuum chamber with replacement of the lens to a concave mirror with the same focal length as the lens. The beam diameter at the focal region can be adjusted with an aperture in front of the focusing optics, resulting in a diameter of the focal spot being tuned approximately from 200 to 400 $\mu$m.

A static gas cell, having entrance and exit pinholes with both diameters of 1.2 mm for the laser beam, in a vacuum chamber is filled with a target gas medium. The gas cell is located near the focal region and the focal position of the fundamental laser field is to be adjusted such that the yield of the harmonic fields should be maximized.

The generated harmonic fields can be measured with an XUV spectrometer placed $\sim$4 m away from the gas cell. The long distance from the focal region ensures the sufficiently low fluence of the laser field to avoid the ablation of a thin metal filter, which is utilized for the calibration of the spectrometer at the wavelength of the absorption edge, inserted in front of the XUV spectrometer.

In order to introduce the harmonic field to an application chamber for spectroscopies, we put a harmonic-separator mirror(s) made of siliconcarbide or silicon,[65] as shown in Fig. 7(b). The reflectivity of the harmonic-separator mirror for the fundamental laser field is measured to be $\sim 10^{-4}$ due to the nature of Brewster incidence of the p-polarized laser field, while that for the harmonic fields is more than approximately 40% in a wavelength range from 26 to 40 nm. The high distinct ratio of the harmonic fields to the fundamental laser field owes to the large difference

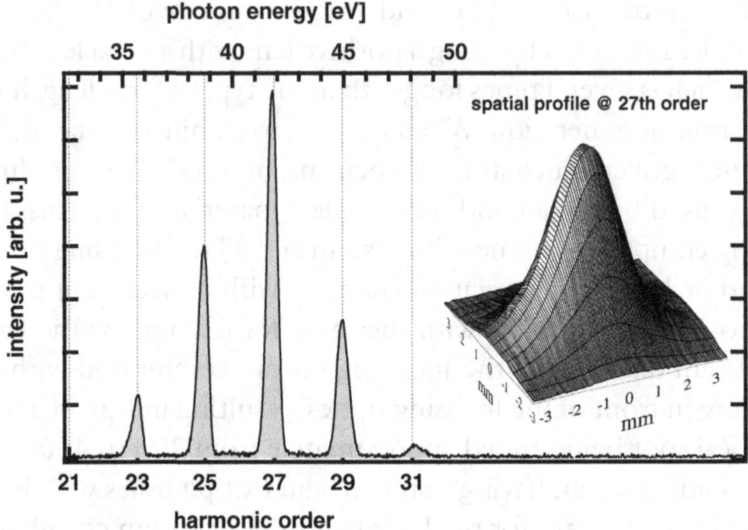


**Fig. 8.** Spectra of high-order harmonic fields generated from the argon gas target. Spatial profile of the 27th order harmonic field is also shown in an inset.

of the refractive index of silicon/siliconcarbide between visible and XUV lights. We can relocate the XUV spectrometer behind the application chamber to monitor the harmonic spectra after reflection from the harmonic separator mirror.

The measured harmonic spectra generated from the argon gas target with a length of 10 cm is shown in Fig. 8.[64] We adjusted the pressure of argon such that the yield of the 27th harmonic field was maximized, resulting in the pressure being 1.8 torr. The spatial profile of the 27th order harmonic field, which was obtained with the transverse scanning of the XUV spectrometer, is also shown in an inset. For this measurement, the pulse energy of the fundamental laser field and the diameter of the aperture are set to 20 mJ and 18 mm, respectively.

We can see the enhancement of the harmonic intensity within the limited range from 25th order to 29th order although the plateau region should be expanded to lower order harmonic fields. This inconsistency comes from the large absorption of the

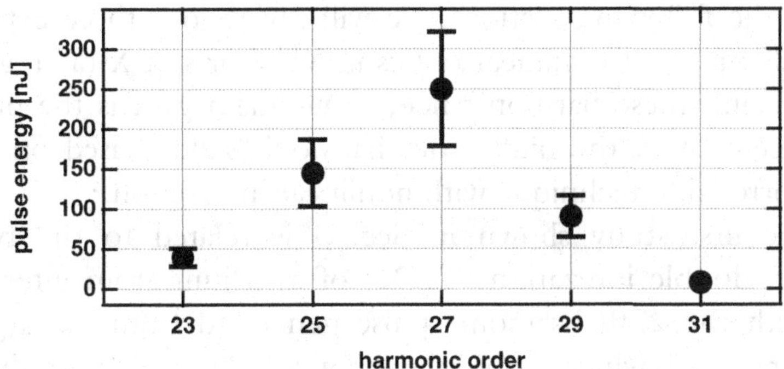


**Fig. 9.** Pulse energy of each harmonic field generated from the argon gas target.

harmonic fields, the orders of which are less than 21st in the argon gas medium. Gaussian-like profile of the harmonic beam with a significantly smaller divergence ($\sim 0.7$ m radians) than that of the fundamental laser field is the evidence of the phase matching.

The pulse energy of the harmonic fields is estimated by measuring the total harmonic yield with an XUV photo-diode is shown in Fig. 9. The highest energy at the 27th order harmonic field exceeds 200 nJ, so that we can expect the intensity to be $10^{12}$–$10^{14}$ W/cm$^2$ for the focused harmonic beam.

We summarize the specifications of the harmonic fields from the three kinds of gas targets in Table 1.[63,64,66,67] The harmonic order, or equivalently the photon energy increases in accordance with the increase of the ionization energy of the gas species while the pulse energy decreases due to the decrease of the induced dipole moment.

**Table 1.**

| | Gas species | | |
|---|---|---|---|
| | Xe | Ar | Ne |
| Principal harmonic orders | 11–15 | 23–29 | 45–61 |
| Photon energy | $\sim$20 eV | $\sim$42 eV | 70–90 eV |
| Pulse energy | 1–5 $\mu$J | $\sim$200 nJ | $\sim$20 nJ |

In the following sections, we will review our three experimental studies of nonlinear optics in XUV or soft X-ray region by utilizing these harmonic fields. Our main goal is the pulse measurement of the high-order harmonic fields based on the autocorrelation technique with nonlinear interaction.

The first study shown in Sec. 3 is related to the two-photon double ionization (TPDI) of a helium atom interacting with the 27th harmonic pulse generated from the argon gas target, of which photon energy is 42 eV. The result of this experiment exhibited the first evidence of a nonlinear phenomenon in the soft X-ray region, and directly revealed the pulse shape of the 27th harmonic field in the sub-10 femtoseconds regime.

In the second study described in Sec. 4, we will introduce two-photon ATI with multiple harmonic fields forming a train of extremely short bunches of the optical pulses, which is conventionally called "attosecond pulse train (APT)." The harmonic fields are obtained from the xenon gas target. The kinetic energy spectra of the electrons with two-photon ATI in the autocorrelation measurement should correspond to the mode-resolved spectra of the APT, so that we can determine the relative phases among the harmonic fields.

Finally, we will report on the first observation of the two-photon Coulomb explosion of a nitrogen molecule, which is utilized for finding the phase-locking and the time-translation symmetry of the APT in the interferometric autocorrelation measurement.

## 3. Two-Photon Double Ionization

Double ionization of a helium atom or some other kinds of rare gas atoms is deduced to reveal a different aspect from the single ionization process because it should contain the correlated properties between the two detached electrons.

The first discovery of double ionization of helium excited by many photons in a visible intense femtosecond laser field appeared as a discrepancy of the doubly-charged-ion yield from the theoretical model based on the sequential ionization from the singly charged ion.[12,30] We can now find the full characteristic of momenta of two electrons leaving a doubly charged ion of neon[73] after many investigations for this phenomenon over a decade. Although there are also many studies for the single-photon double ionization of helium,[21,28,74] which is the simplest process producing a correlated electron pair with a photon, by using the soft X-ray light from synchrotron sources, so far we did not have the experimental proof of the second simplest process; the two photon double ionization (TPDI) of helium.

Fortunately, we have the intense ultrafast light source of the 27th order harmonic field with a photon energy of 42 eV generated from the argon gas target, as described in the previous section, which is suitable for observing TPDI in a helium atom, whilst the physical model of production of a doubly charged ion of helium with this photon should include three prime pathways for the nonlinear process for ionization,[18] as schematically shown in Fig. 10. The first of the three pathways is sequential ionization from a singly charged ion to a doubly charged one through two photon absorption. The photon energy of the 27th harmonic pulse exceeds the ionization energy from the neutral atom to the singly charged ion, whereas the resulting singly charged ion, needs two photons to release the rest of the electron from the nuclei in order to climb the 54-eV potential wall, as is indicated by solid arrows at the right-hand side in Fig. 10. Thus, three photons are needed to fully ionize the neutral atom of helium in this process.

The second pathway requires three or four photons via the excited state or the ground state of an electron on a singly charged ion which is produced by the two-photon absorption of a neutral atom, as shown at the center in Fig. 10. The excess of the energy

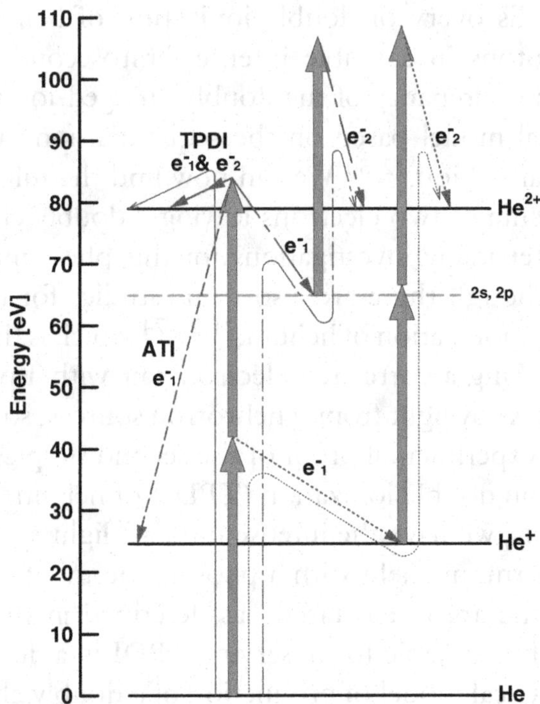


**Fig. 10.** Energy diagram of helium. The thick solid arrows correspond to the photon energy of the 27th-order harmonic field. Detached electrons are symbolized as $e_1^-$ and $e_2^-$. Thin dotted arrow indicate a sequential double ionization process via the ground state of a singly charged helium ion. The doubly charged ion can also be generated along the pathway depicted with a thin dot-dashed arrow. The singly charged ion should be excited with two-photon absorption, then further ionized with one-photon absorption in this pathway. Other excited states of the singly charged ion can contribute this kind of ionization process. The TPDI process is shown as a thin solid arrow.

of photon to that of the bounded electron in the singly charged ion should be transferred to the kinetic energy of the released electron. Therefore, this process contains ATI.

The third pathway is simply TPDI from a neutral atom to a doubly charged ion without passing through the any states for a singly charged ion. We note that this nonlinear process cannot be induced until the photon energy exceeds half of the energy required for double ionization, namely, 39.5 eV, which is much higher than the photon energies applied in the nonlinear experiments of high-order harmonics reported so far.[29,31,40,59,60,70]

These three processes can be distinguished with an electron spectrometer by observing the kinetic energy of each electron via each ionization pathway, in principle. It is, however, not easy to find these electrons in practice, because the huge amount of electrons originating from the one-photon ionization of helium atom disturbs the correct observation of the extremely small signal of these electrons. In particular, a pair of electrons via TPDI is expected to distribute its kinetic energy into the wide range of the spectrum from 0 to $\sim 5$ eV, and consequently, the counted numbers of the electrons per unit energy should decrease below the signal to noise ratio. The only nonlinear phenomenon of the 27th order harmonic field that we could find by observing the electrons was the two-photon ATI.[18] Our success of the experiment owes to the fact that the two-photon ATI peak in the electron spectra is far from the one-photon ionization peak in a time-of-flight measurement, while this phenomenon is classified as a single ionization process.

Thus, we measured the time-of-flight mass spectra of ions instead of electrons.[18,41] Although we cannot exactly determined how the three processes contribute the yield of the doubly charged helium ion, we can acknowledge the observation of this ion yield itself as clear evidence of the nonlinear phenomenon by considering that any of the three processes requires two or more photons.

The experimental setup for the ion spectroscopy is shown in Fig. 11. The reflected high-order harmonic fields at all orders with a harmonic separator made of siliconcarbide are sent into a vacuum chamber for the measurement of the time-of-flight of ions for which they are passed through a 1-mm diameter hole bored at the center of two thin plates of stainless steel for acceleration.

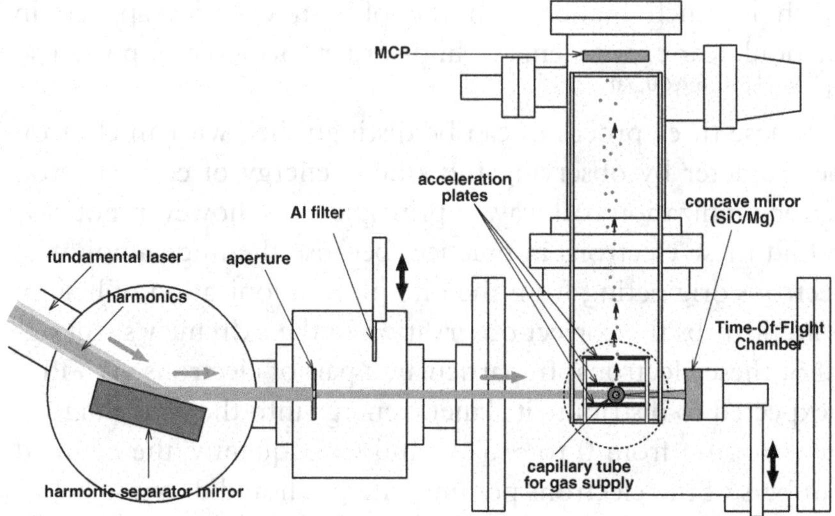


**Fig. 11.** Experimental setup of time-of-flight mass spectroscopy for the detection of the doubly charged helium ions.

The beam of the harmonics passes 5 mm away from the center of the 1 mm hole in the acceleration plates in order to prevent a microchannel plate (MCP), which is placed 385 mm away from the hole, from detecting ions produced with the unfocused beam. A multi-layer coat of siliconcarbide and magnesium on the surface of a concave substrate made of fused silica with a radius of curvature of 100 mm reflects 24% of the 42 eV photons with a bandwidth of 2.2 eV, which means that almost only the 27th harmonic field in the harmonics can be reflected by this concave mirror, and is focused in front of the hole. The spectrum of the harmonics can be monitored with a spectrometer set behind the time-of-flight chamber, by removing the concave mirror. We truncate the undesired edge of the harmonic beam by placing an aperture with a diameter of 3 mm between the beam separator and the time-of-flight chamber, while the central part of the harmonic beam including approximately 70% of the harmonic energy can pass through this aperture due to the small divergence of the phase-matched harmonic field. This aperture also eliminates ~94% of the energy of the remaining fundamental pulse reflected by the harmonic separator, with the result of that the energy of the fundamental pulse at the focal point of the concave mirror is reduced to ~ 100 nJ from 20 mJ at the entrance of the 5 m focal lens for the harmonic generation.

The intensity of the 27th harmonic at the focal point, that we notate as $I_0$ in this section, is estimated to be $1.7 \times 10^{13}$ W/cm$^2$, assuming a 3 $\mu$m beam waist ($w_0$),[67] a 10 fs pulse duration ($\tau_p$), and a 24 nJ pulse energy ($E_0$). The intensity of the remaining fundamental laser field is estimated to be $\sim 2 \times 10^{12}$ W/cm$^2$, which is approximately $10^{-2}$ of the intensity at which the helium ion can appear for the visible laser pulse.[72] In fact, we did not observe any helium ions when we applied only the fundamental pulses by switching off the supply of argon gas as a nonlinear medium for the harmonic generation.

The signals of ions detected with the MCP are sent to a digital oscilloscope with a bandwidth of 500 MHz and a sampling rate of 5 Gs/s, and then they are typically averaged over $10^3$ shots for each measurement. We note that the signal of the doubly charged helium ion, is sufficiently large to appear in the averaged detection without the signal counting method conventionally used in this type of experiment.

Helium gas, as a target of the ion signals, is supplied via a glass tube with an inner diameter of 800 $\mu$m, which is connected to a pulsed gas valve operated synchronously with the laser pulse. We use the isotope helium 3 instead of helium 4 because our setup for the time-of-flight experiment cannot resolve the signal of the singly charged ion of a hydrogen molecule from that of the doubly charged ion of a helium 4 atom. The estimated density of the atoms of helium 3 is $\sim 3 \times 10^{13}$ cm$^{-3}$.

A typical mass spectrum for the doubly charged ions of helium 3 is shown in Fig. 12.[18,41] We can clearly see the presence of a peak at the mass number of 1.5 corresponding to the half of

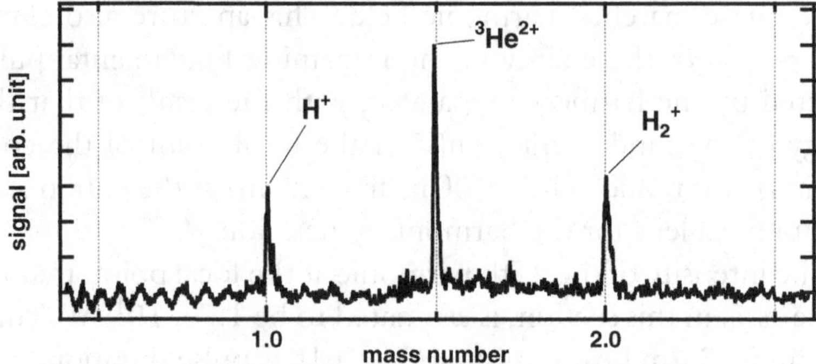


**Fig. 12.** Typical mass spectrum of doubly charged ions of helium 3 generated with the 27th harmonic field of which photon energy is 42 eV.

the mass for the helium 3 atom. The singly charged ions of the hydrogen atom at 1.0 and those of the hydrogen molecule at 2.0 mainly originate from water molecules that remain in the TOF chamber at the background pressure of $8 \times 10^{-8}$ torr.

The presence of the doubly charged helium ions ensures, at least, the existence of a nonlinear process as described in this section. We can say that this experimental result is the first observation of the nonlinear process in the soft X-ray region. We do not know, however, which of the three pathways dominantly contributes for this ion yield. Thus, we measured how the ion yield was affected by change of the intensity of the harmonic field to qualitatively find the most dominant process of the three.

In order to simultaneously measure the relative intensities of the 27th harmonic pulse and the corresponding yields of doubly charged helium ions at those intensities, we utilize the yields of singly charged helium ions as indicators of the intensities. We can assume that the yield of the singly charged ions is proportional to the intensity of the 27th harmonic because there is no saturation in their production. The probability of the ionization at the focal point, notated as $P_0^{(+)} = I_0 \tau_p \sigma^{(+)}/(h\nu)$, can be estimated to be much lower than unity $(7 \times 10^{-2})$ even if the intensity is the maximum,[54] where $\sigma^{(+)}$, $\nu$, and $h$ are the cross-section of one-photon ionization, the optical frequency of the 27th harmonic, and Planck's constant, respectively.

A gated boxcar integrator, which are terminated with a channel of the digital oscilloscope behind the integrator, receives the electrical signal from the MCP and it integrates a gated signal of the singly charged ion to each laser shot. The integrated signals sent to the channel of the oscilloscope are recorded with statistical fluctuations, while the TOF spectra in another channel of the oscilloscope including the signal of doubly charged ion are averaged to $10^3$ shots of the laser pulse for each measurement at the fixed intensity of the harmonic field. The intensity of the 27th harmonic pulse is adjusted by changing the energy of fundamental pulse.

The result of this experiment is shown in Fig. 13. The yields of the doubly charged ions in relation to those of the corresponding singly charged ions are well located on the fitted line of $N^{(2+)} = A(N^{(+)})^{\alpha}$ in the logarithmic scale for both axes in

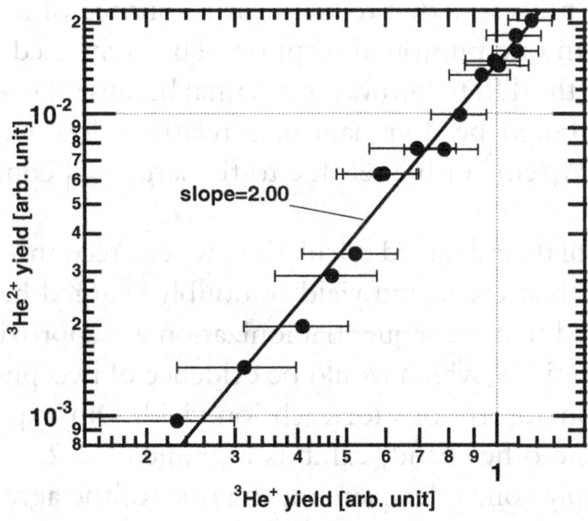


**Fig. 13.** Yields of the doubly charged ion of helium 3 versus those of the singly charged ion. Both are varied parametrically with the intensity of the 27th harmonic pulse.

this figure, where $N^{(2+)}$ is the yield of the doubly charged ion, $N^{(+)}$ is the yield of the singly charged ion, and $A$ and $\alpha$ are the constants determined by fitting the line to the experimental data, respectively. We conclude that the exponent $\alpha$ is 2.0, and hence the two-photon absorption with double ionization is found to be the most dominant process in this experiment.

It is important to compare an experimental result with a theoretical prediction for understanding of nature.[6,10,13,32,39,45,49] Nikolopoulos and Lambropoulos have performed detailed quantitative calculations of the yields of singly charged helium ion, doubly charged ion via sequential ionization from the singly charged ion, and doubly charged ion via TPDI, respectively,[48,50] under the similar condition to our experiment. They conclude that the nonlinear cross section of the TPDI process, $\sigma_{\mathrm{TPDI}}$ should be $0.4 \times 10^{-52}$ cm$^4$ s, and the nonlinear cross section of the two-photon ionization from a singly charged ion to a doubly charged ion, $\sigma_{\mathrm{seq}}$, should be 4.5 times of that of the TPDI, namely, $1.8 \times 10^{-52}$ cm$^4$. The larger nonlinear cross section of this process is due to the near resonance of the 2p and 2s states with one-photon absorption. They criticized our conclusion that the TPDI pathway is essential because the sequential ionization should be dominant at a relatively low intensity of $9.2 \times 10^{13}$ W/cm$^2$ or higher due to the large $\sigma_{\mathrm{seq}}$ compared to $\sigma_{\mathrm{TPDI}}$.

In spite of their detailed calculation, we can recognize that the gradient of their calculated yield of doubly charged helium ion via TPDI and that via sequential ionization are approximately 2 and 3, respectively, which would be evidence of two-photon and three-photon absorptions for each ion yield. Our experimental result, on the other hand, exhibits a gradient of 2, and therefore we simply conclude again, on the basis of the agreement of the gradient of the fitted line with that of the calculated curve, that TPDI was the dominating process and that most of all the intensities at the data points in Fig. 13 had to be sufficiently lower than the critical intensity of $9.2 \times 10^{13}$ W/cm$^2$.[42]

As we have confirmed the nonlinearity of the yield of the doubly charged helium ion, we can utilize this ion yield for the measurement of the pulse duration of the 27th harmonic pulse by means of an autocorrelation technique. Two replicas of the measured pulse are generally needed for the autocorrelation measurement as was explained in Sec. 1.2. However, we do not have any substrate materials suitable for a half mirror that are highly transparent to XUV or soft X-ray light. We solve this issue by splitting the spatial profile of the measured harmonic field using split harmonic separator mirrors placed as closely to each other as possible.

The center of the harmonic beam enters the boundary of the separator mirrors, and then it is reflected separately. This spatial division of the harmonic beam is confirmed with a MCP and a CCD camera attached to the spectrometer behind the time-of-flight chamber. We observe the semicircle profiles of the divided harmonic beams.

One of the harmonic separators is fixed on a translation stage with a piezoactuator that can move the stage within a length of 100 $\mu$m, hence the part of the harmonics reflected with this harmonic separator can be delayed or advanced to another part reflected with another harmonic separator within a range of $\pm$120 fs, which is calculated from the incident angle common to both harmonic separators of 69 degrees corresponding to the Brewster incident angle for SiC substrate.

The yield of doubly charged helium ions to each delay is shown in Fig. 14. We can see a typical autocorrelation trace near zero delay in this figure while the pedestals, which are inherited from those of the fundamental pulse, appear at a delay of $\pm$ 40 fs. As was calculated by Tzallas *et al.*,[70] the ratio of the peak height to the background in the autocorrelation signal integrated to a certain volume of the interaction region should be approximately 3, which agrees well with the ratio in the trace in Fig. 14. We note that this experimental result of the autocorrelation is supported by the theoretical investigation reported by Nakajima *et al.*[45]

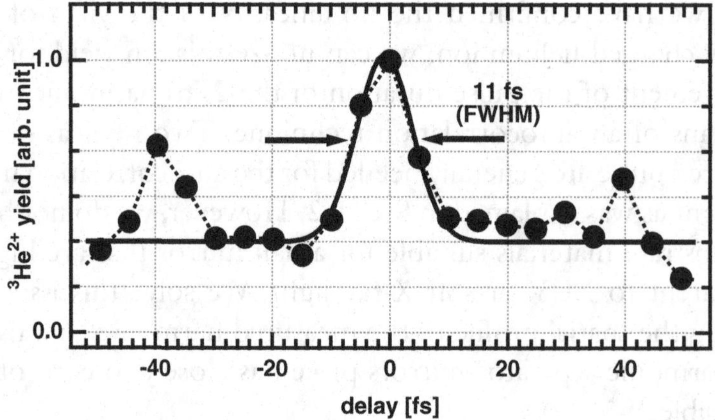


**Fig. 14.** Autocorrelation trace (solid circles with dotted line) of the 27th harmonic pulse obtained by utilizing the two-photon absorption for the yield of a doubly charged ion of helium 3. The full width at half maximum (FWHM) of the autocorrelation is determined by fitting a Gaussian shape (solid curve) to the trace.

A curve fitted to the central region of the trace assuming a Gaussian temporal profile, indicated as a solid curve in Fig. 14, results in a pulse duration of 8 fs corresponding to $1/\sqrt{2}$ of the full width at half maximum of the fitted curve. It is difficult to conclude whether the pulse duration of the 27th harmonic pulse is reasonable or not with respect to the SFA model of harmonic generation described in Sec. 2.1 because this model assumes the continuous electric field as a fundamental laser. Numerical calculation for solving the Schrödinger equation combined with the propagation equation based on the Maxwell equation may be needed to solve this issue.

## 4. Measurement of Attosecond Pulse Train with Two-Photon ATI

In the experiment described in the previous section, we selected one of the harmonic fields in the plateau region with a bandwidth-limited reflection of the multilayered coat on the concave mirror. Thus, the temporal profile of the harmonic field is a single-pulse shape reflecting the intensity profile of the fundamental laser field although the pulse duration was shortened. Then how does the synthesized field of the multiple order harmonics appear in time domain? We have already known, from the description of Sec. 2.2, that the spectral phase of each harmonic field is uniquely specified with the phase matching effect. Thus, the synthesized electric field should be simply modeled as the coherent summation of the harmonic fields.[1]

$$E^{+}(t) = \sum_{n=n_1}^{n_2} A_{2n+1} e^{-i(\omega_{2n+1}t+\phi_{2n+1})}, \tag{8}$$

where $E^{+}(t)$ is the positive frequency part of the synthesized electric field of the harmonics at time $t$. The real amplitude and the phase of the $(2n+1)$th harmonic field are notated as $A_{2n+1}$ and $\phi_{2n+1}$, respectively. The harmonic orders are restricted within a range of $[2n_1+1, 2n_2+1]$ in this model. By considering that the angular frequency of the $(2n+1)$th order harmoic field, $\omega_{2n+1}$, is equal to $(2n+1)$ times the angular frequency of the fundamental laser field, which is notated as $\omega_f$ in the following equations, we can recognize that Eq. (8) represents a kind of Fourier synthesis of equally separated modes of waves, of which mode-separation is $2\omega_f$. Although the finite frequency offset (FO) at $n = 0$ disturbs the complete time-translation symmetry of $E^{+}(t)$ as is satisfied in an usual Fourier synthesis, the intensity $I(t)$, which is proportional to the absolute square of $E^{+}(t)$, is exactly periodic with a time period of $T_f/2$, where we define $T_f \equiv 1/\nu_f \equiv 2\pi/\omega_f$.

In particular, the intensity profile exhibits very short bunches if the phases of the modes are not considerably differed each other. Let us assume all the amplitude and the phase are equal to $A$ and $\phi$, respectively, in order to see this situation. The right-hand side of Eq. (8) can be easily summed up, then we obtain the synthesized electric field as

$$E^{+}(t) = A\frac{\sin(N\omega_{f}t)}{\sin(\omega_{f}t)}e^{-i(\omega_{f}\overline{n}t+\phi)}, \tag{9}$$

where $N$ is the number of modes, namely, $N \equiv n_2 - n_1 + 1$. Thus, the intensity profile should be proportional to $\sin^2(N\omega_f t)/\sin^2(\omega_f t)$ having its maximum at $t = qT_f/2$ ($q = 0, \pm 1, \pm 2, \ldots$) and forming the sequence of pulses with a width of approximately $1/(N\nu_f)$. The carrier frequency of this pulse sequence should be the average of the frequencies because the number $\overline{n}$ is equal to $n_1+n_2+1$ $(=\{(2n_1+1)+(2n_2+1)\}/2)$. The optical frequency of the fundamental laser field is typically $\sim$375 THz, so that the width of each pulse should be much shorter than 1 fs even when the only three harmonic fields participate to form the synthesized electric field. This type of sequence of extremely short bunches formed with high-order harmonic fields is customary called "attosecond pulse train" (APT).[1] The period of the pulses in the APT is $\sim$1.33 fs for the fundamental laser field from a Ti:sapphire laser system.

We neglect the bandwidth of each harmonic field in Eq. (8), resulting in the pulse train infinitely continuing in time domain, while the APT envelope, in actual, continues within a limited duration of $\sim$ 10-fs range depending on the pulse duration of the fundamental laser field. Thus, we should note that the simple model shown as Eq. (8) is a good approximation only for the central part of the APT envelope at which the peak intensity of each pulse dose not notably change. We now treat the issue of the train nature of the synthesized harmonic field arising from the sinusoidal wave form of the fundamental laser "field" itself with an exact periodicity. The measured pulse shape of the one harmonic field shown in the previous section, in contrast, is related to the APT envelope originating from the "intensity" profile of the laser field.

The APT formed with high-order harmonic fields have been investigated with a kind of cross correlation techniques, called RABITT, which is an acronym of "reconstruction of attosecond beating by interference of two-photon transition."[52] The sidebands induced with two-color ATI appear in the comb-like photoelectron spectra when a rare gas (argon) atom is simultaneously irradiated with the fundamental laser field and the multiple harmonic fields. The electron yield in the $2n$th sideband comes from both the electrons absorbing one photon of the $(2n-1)$th harmonic field with one more photon absorption of the fundamental laser field, while it is also contributed with the electrons absorbing one photon of the $(2n+1)$th harmonic field with one more photon emission of the fundamental laser field within the approximation of the second order perturbation for the transition amplitude of the electron. The sinusoidal modulations of the sidebands, with a period of $T_f/2$, should be found by scanning delay between the harmonic field and the fundamental laser field, because of the interference between the fundamental laser field for the absorption and the conjugated field for the emission, both of which contribute coherently to the transition amplitude. The phase of the modulation at the $2n$th sideband is shifted by $\phi_{2n+1} - \phi_{2n-1}$ due to the similar interference between the $(2n-1)$th and $(2n+1)$th harmonic fields in the transition amplitude, and therefore we can extract the phase difference of two adjacent harmonic fields by observing this phase shift of each side band, then, the temporal shape of the synthesized harmonic field can be reconstructed by calculating Fourier synthesis of the measured spectral amplitudes of harmonic fields with the phases obtained from this method. This is the RABITT.

Although the RABITT is feasible to find the APT nature of the high-order harmonic fields, it contains drawbacks for the pulse measurement. The phase shift of the sideband does not only arise from the phase difference of the adjacent harmonic fields but also originates from the dipole transition matrix elements via all the intermediated states. In order to extract the correct phases from the experimental data, we have to know all the correct phases of these matrix elements. Paul *et al.* numerically calculated this phase factor for each sideband[52,69] by assuming the effective potential of argon that can reproduce the excited states of the actual argon atom. The integration of the wave function of the electron with respect to the solid angle coordinates can be significantly reduced to the summation of a few quantum numbers for the angular momentum because the selection rules are rigorously correct due to the spherical symmetry of the effective potential. On the other hand, the numerical integration method for the radial wave functions is too technical for nonspecialists of the theory of atomic physics. Further, the phenomenon of the sideband generation is not classified into the nonlinear optics of XUV/soft X-ray light itself, which is the main concern of this article. Thus, we will concentrate our attention into the autocorrelation measurement of the APT by utilizing the nonlinear optics with an interaction of two photons both of which photon energies are categorized into the XUV/soft X-ray region.

Two-photon absorption (TPA) in a rare gas atom might be one of the most favorable nonlinear phenomena for measuring the APT, because it should be a simpler process than that in molecules having degrees of freedom of the nuclear motion other than the electron motion. The first observation of an TPA in an atom induced by the XUV harmonic field of a femtosecond Ti:sapphire laser was reported by Kobayashi *et al.*[29] as was mentioned in Sec. 1.2. Tzallas *et al.* found the autocorrelation trace of an APT in a similar manner.[70] They have shown the single-ionization of a helium atom induced with TPA of the multiple harmonic fields ranging from 7th to 15th orders, and then demonstrated the modulation of the ion yields with a period of $T_f/2$ in their autocorrelation measurement. The pulse width estimated in their experiment have been much longer than that expected.

The autocorrelation trace by the observation of the ion yields, however, does not give us an enough information for determining the pulse shape of an APT because the spectral-phase difference between the harmonic fields cannot be found in that trace. In order to fully characterize the intensity profile of the APT, we have to resolve spectrally the autocorrelation trace, as is conventionally carried out for measuring the femtosecond laser pulse in the frequency resolved optical gating (FROG) method.[25] The electron spectra originating from two-photon above-threshold ionization (TP-ATI) can be used for the spectrally resolved autocorrelation. We have already shown the schematic figure of the energy diagram of the TP-ATI in Fig. 10, as a thick dashed line with an arrow.

The electron ionized with one-photon absorption absorbs one more photon, resulting in the TP-ATI electron. Thus, we can identify the TP-ATI electrons, the kinetic energy of which corresponds to twice the photon energy with subtraction of the ionization energy, by observing the electron spectra.

Hereafter in this section, we introduce our study of the autocorrelation measurement of the APT consisting of the three harmonic fields at the 11th, 13th, and 15th orders by observing the electrons via TP-ATI of an argon atom.[16,43] The experimental result clearly indicates the formation of an APT at a glance. The analysis of the TP-ATI spectra, which can be considerably simplified due to the small numbers of harmonic fields, specifies that the chirp (group delay dispersion) among the three harmonic fields should be smaller than $1.4 \times 10^{-32}\ s^{-2}$, resulting in a pulse duration of 450 attoseconds.

The setup of the experiment is similar to that explained in the previous section. A fundamental laser pulse with a duration of 40 fs and an energy of 13 mJ is delivered from a chirped pulse amplification (CPA) system of a Ti:sapphire laser, and focused into a static gas cell filled with xenon gas, instead of argon gas for the efficient harmonic yield ranging from 11th to 15th orders, in a vacuum chamber. The pulse energies of the 11th-, 13th-, and 15th-order harmonic fields just behind the gas cell are all estimated to be higher than 1 $\mu$J,[63,67] and these high pulse energies are critical for observing the nonlinear interaction of the harmonic field without the help of another intense laser field. The generated harmonic fields and the fundamental laser copropagate to the next vacuum chamber including a pair of harmonic separators made of silicon.[65] The configuration of the harmonic-separator mirrors for making two replicas of the harmonic fields is the same as that used in the previous experiment in Sec. 3. except for the incident angle being tuned 75 degrees in accordance with the refractive index of silicon.

The remaining portion of the fundamental laser, typically $\sim 10^{-4}$ of the incident pulse energy, and the harmonic fields with an order lower than the 9th are eliminated with a tin filter with a thickness of 0.1 $\mu$m, of which the measured transmittances of the 11th-, 13th-, and 15th-order harmonic fields are 6%, 14%, and 20%, respectively. We truncate the undesired edge of the harmonic beam by placing an aperture with a diameter of 3 mm in front of the tin filter.

Two replicas of the harmonic fields are introduced in a magnetic bottle photoelectron spectrometer and focused with a concave mirror made of silicon carbide with a radius of curvature of 100 mm. Argon gas, as target atoms of two-photon ATI, is supplied with a capillary tube attached to a pulsed gas valve. Approximately half of the ejected electrons with photoionizations are guided with a magnetic field, and then go through a small bore with a radius of 700 $\mu$m. The kinetic energies of the electrons are resolved by measuring the time of flight of the electrons traveling into a flight tube behind the bore with the guidance of the magnetic field.

The harmonic spectra at the focal region of the concave mirror were determined by observing electron spectra with the one-photon absorption of an argon atom because the photon energies of the 11th- and higher-order harmonic fields exceed the ionization energy of the argon atom. The resultant relative intensity ratios of the 11th-, 13th-, and 15th-order harmonic fields, which we, respectively, denote as $I_{11}$, $I_{13}$, and $I_{15}$ in this section, are 0.46:1:0.28, and the intensities of higher-order harmonic fields are negligible compared with $I_{13}$ due to the tin filter, but measurably large compared with the TP-ATI signals up to the 23rd order, as is described later. The reductions of the 9th- and 7th-order harmonic fields were confirmed using a VUV–XUV spectrometer.

The possible kinetic energies of electrons corresponding to the two-photon ATI peaks with these three harmonic fields are schematically shown in Fig. 15. The kinetic energies of the electrons with the offset of the ionization potential are given in units of photon energy of the fundamental laser field for convenience, where we define $\Omega_{2n}$ as $2n\omega_f$. We can expect five peaks, from the lowest-order mode of $\hbar\Omega_{22}$ originating from two photons of the 11th-order harmonic field, to the highest-order mode of $\hbar\Omega_{30}$ originating from two photons of the 15th-order harmonic field.

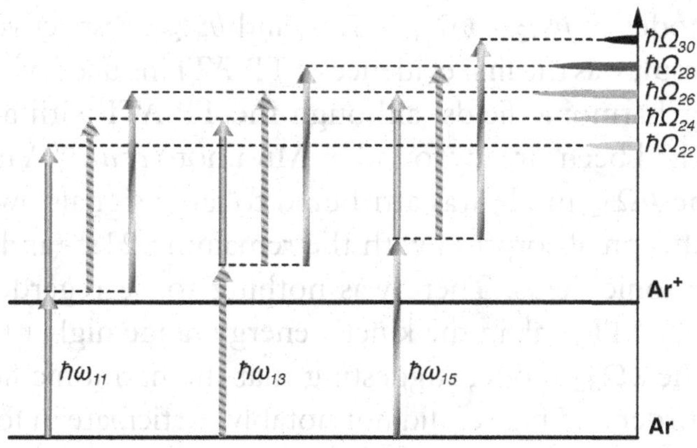


**Fig. 15.** Schematic figure of energy diagram of TP-ATI induced with multiple harmonic fields. A TP-ATI peak of the electron spectrum is emerged through multiple pathways interacting with two photons.

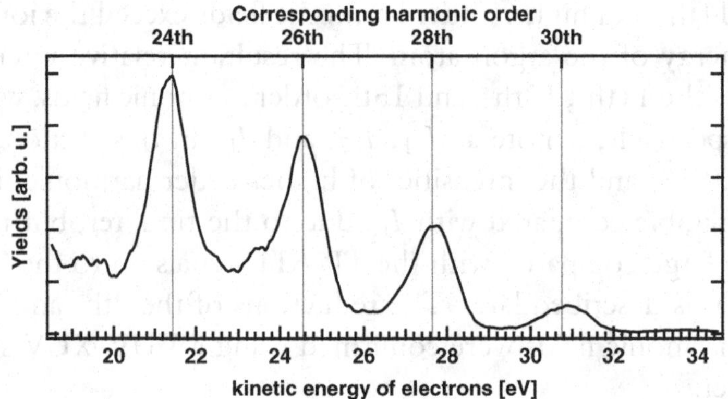


**Fig. 16.** TP-ATI spectra observed with an electron spectrometer. These four peaks corresponds to the two-photon modes of the APT.

We have successfully observed TP-ATI spectra at a fixed delay of the two harmonic-separator mirrors, shown in Fig. 16. The peak of the energy spectrum at the most left in this figure is assigned to 21.4–21.5 eV, which corresponds to the energy subtracting the ionization potential of an argon atom, 15.76 eV, from the energy of the 24th two-photon mode, 37.2 eV. The each energy separation between adjacent peaks corresponds to $2\,\hbar\omega_{\mathrm{f}}$, thus we can conclude that these four peaks are equivalent to two-photon modes of $\hbar\Omega_{24}$, $\hbar\Omega_{26}$, $\hbar\Omega_{28}$, and $\hbar\Omega_{30}$, respectively.

This result was the first evidence of TP-ATI induced with multiple order harmonic fields, although the TP-ATI with a single harmonic had been already found by Miyamoto *et al.*[40] A missing peak in the $\hbar\Omega_{22}$ mode was attributed to large signals owing to the one-photon absorption with the remaining 21st- and 23rd-order harmonic fields. There was nothing to be regarded as a two-photon ATI peak in the kinetic energy range higher than or equal to the $\hbar\Omega_{32}$ mode, suggesting that the harmonic fields of the 17th-order and higher did not notably participate in forming the attosecond pulse train. This result confirms the assumption that there are only three harmonic fields in our analysis described later.

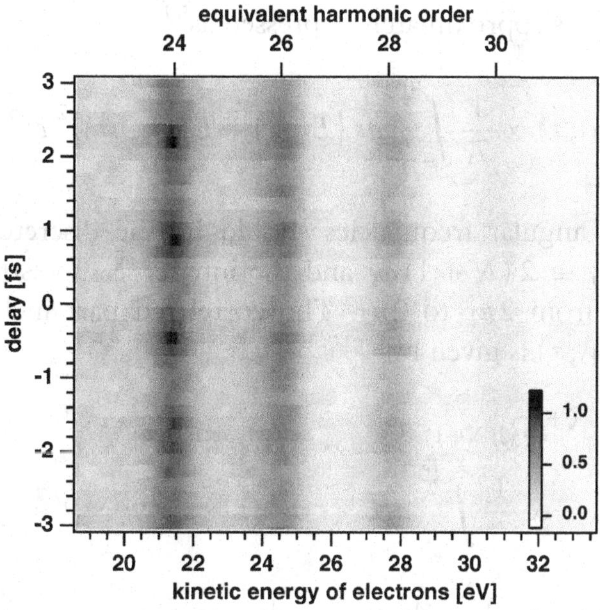


**Fig. 17.** Energy spectra of TP-ATI electrons depending on delay. This image plot corresponds to the mode-resolved autocorrelation of the synthesized harmonic field. The clear modulations at the 24th, 26th, and 28th orders with the same period of 1.33 fs, which is in agreement with the half period of the fundamental laser field, are both theoretically and intuitively recognized as evidence of the formation of the attosecond pulse train.

By changing the delay between the two replicas of the harmonic fields, these four TP-ATI peaks should reveal the correlated signal of the temporal profile synthesized with the harmonic fields to each energy mode. In fact, we clearly observed the modulations of the TP-ATI peaks with a period of $T_f/2$ (1.33 fs) in the $\hbar\Omega_{24}$, $\hbar\Omega_{26}$, and $\hbar\Omega_{28}$ modes, as shown in the image plot of Fig. 17.

Before saying eureka, we should briefly analyze how the attosecond pulse train causes this mode-resolved autocorrelation. We have already defined the positive frequency part of the synthesized harmonic field in Eq. (8). With this expression, the Fourier amplitude of the signal, $\tilde{S}^{+}_{\text{all}}(\Omega_{2(N+1)};\tau)$, with the nonresonant two-photon process induced by two replica of the electric field with delay $\tau$ is approximately expressed as[59]

$$\widetilde{S}^{+}_{\text{all}}(\Omega_{2(N+1)};\tau) \propto \frac{1}{T_{\text{f}}}\int_{-\frac{T_{\text{f}}}{2}}^{\frac{T_{\text{f}}}{2}} dt \left\{E^{+}(t)+E^{+}(t-\tau)\right\}^{2} e^{i\Omega_{2(N+1)}t}, \tag{10}$$

where the angular frequencies should be the discrete modes of $\Omega_{2(N+1)} \equiv 2(N+1)\omega_{\text{f}}$ and the integer $N$ is restricted in the range from $2n_1$ to $2n_2$. The correlated part included in $\widetilde{S}^{+}_{\text{all}}(\Omega_{2(N+1)};\tau)$ is given by

$$\begin{aligned} \widetilde{S}^{+}&(\Omega_{2(N+1)};\tau) \\ &\propto \frac{1}{T_{\text{f}}}\int_{\frac{-T_{\text{f}}}{2}}^{\frac{T_{\text{f}}}{2}} dt E^{+}(t)E^{+}(t-\tau)e^{i\Omega_{2(N+1)}t}, && (11) \\ &= \sum_{n=n_{\min}}^{n_{\max}} A_{2(N-n)+1}A_{2n+1} \\ &\quad \times e^{-i(\phi_{2(N-n)+1}+\phi_{2n+1})} e^{i(2n+1)\omega_{\text{f}}\tau}, && (12) \end{aligned}$$

where the pair of integers $(n_{\min}, n_{\max})$ should be equal to $(n_1, N-n_1)$ for the case of $N \leq n_1+n_2$ and $(N-n_2, n_2)$ for $N > n_1+n_2$. This equation simply reflects the convolution of the Fourier amplitude of each electric field within a finite range of modes. The observed electron spectra should reflect the absolute square of $\widetilde{S}^{+}(\Omega_{2(N+1)};\tau)$.

The electron yields for the actual experiment is approximately proportional to $|\widetilde{S}^{+}_{\text{all}}(\Omega_{2(N+1)};\tau)|^2$, so that they include a constant background originating from $|\widetilde{S}^{+}(\Omega_{2(N+1)};0)|^2$, and the oscillating terms corresponding to the carrier frequencies of the harmonic fields, which originate from $\widetilde{S}^{+}(\Omega_{2(N+1)};0)\widetilde{S}^{+*}(\Omega_{2(N+1)};\tau)$. These are also affected by the volume integration in the focal region. The oscillating terms could not be resolved in the experiment because the resolution of the delay in the experiment is limited to 165 attoseconds, which is not sufficient for finding the interference fringes. Further, the inhomogeneous thickness of the tin filter would significantly degrade the visibility of the spatial fringes at the focal region of the harmonic field, which is essential for the interferometric autocorrelation by utilizing the spatially separated two replicas of the measured field, even if the resolution of the delay had been adequate.

The odd integers of $2n_1 + 1$ and $2n_2 + 1$ are 11 and 15, respectively, in our experiment, thus, the mode-resolved electron yielding $I(\Omega_{2(N+1)};\tau) \propto |\widetilde{S}^{+}(\Omega_{2(N+1)};\tau)|^2$ can be written as $I(\Omega_{22};\tau) \propto I_{11}^2$, $I(\Omega_{30};\tau) \propto I_{15}^2$, and

$$I(\Omega_{24};\tau) \propto 2I_{11}I_{13}\{1 + \cos(2\omega_f\tau)\}, \tag{13}$$

$$\begin{aligned} I(\Omega_{26};\tau) \propto\ & \{I_{13} + 2\sqrt{I_{11}I_{15}}\cos(2\omega_f\tau)\}^2 \\ & - 8I_{13}\sqrt{I_{11}I_{15}}\sin^2(\Delta\Phi)\cos(2\omega_f\tau), \end{aligned} \tag{14}$$

$$I(\Omega_{28};\tau) \propto 2I_{13}I_{15}\{1 + \cos(2\omega_f\tau)\}, \tag{15}$$

where we define the intensity of each harmonic field as $I_{2n+1} \equiv \beta A_{2n+1}^2$ $(2n + 1 = 11, 13, 15)$ with an appropriate constant, $\beta$, and the phase difference $\Delta\Phi \equiv (\phi_{11} + \phi_{15})/2 - \phi_{13}$.

Trivial equations for the 22nd- and 30th-order modes can be easily recognized as results that only one harmonic field is concerned in generating these modes. The two adjacent harmonic fields responsible for generating the 24th- and 28th-order modes cause a sinusoidal modulation with the half period of the fundamental laser field as we have expected. Note, however, that the spectral phases of the harmonic fields, $\phi_{11}$, $\phi_{13}$, and $\phi_{15}$, disappear in these equations, so that we cannot extract any relations among the spectral phases of the three fields from the measured autocorrelation traces in these two modes, while these two modes should always exhibit the perfect sinusoidal modulation whatever happens to the spectral phases. Thus, these two modes give us benchmarks for determining the correlated part of the ATI signals on the constant background.

The last mode we should consider is the 26th-order mode. The intensity of $I(\Omega_{26};\tau)$ is modulated with mainly the half period of the fundamental laser field because of the first term in Eq. (14), and somewhat corrected with the second term proportional to $\sin^2(\Delta\Phi)$, which includes information on the relative phases of the three harmonic fields, so that we should be able to determine $|\Delta\Phi|$ from the measured profile of $I(\Omega_{26};\tau)$.

We have to consider what $\Delta\Phi$ is and how it affects the temporal profile of the pulse train before determining it. The general expression of the intensity of the synthesized harmonic field, denoted $I(t)$, can be written from the absolute square of Eq. (8) as follows.

$$\begin{aligned} I(t) = I_{11} + I_{13} + I_{15} + 2\sqrt{I_{11}I_{13}}\cos\{2\omega_{\mathrm{f}}(t+\tau_{\mathrm{g}}) - \Delta\Phi\} \\ + 2\sqrt{I_{13}I_{15}}\cos\{2\omega_{\mathrm{f}}(t+\tau_{\mathrm{g}}) + \Delta\Phi\} \\ + 2\sqrt{I_{15}I_{11}}\cos\{4\omega_{\mathrm{f}}(t+\tau_{\mathrm{g}})\}. \end{aligned} \tag{16}$$

Here, we define $\tau_{\mathrm{g}} \equiv (\phi_{15} - \phi_{11})/(4\omega_f)$.

Obviously, the parameter $\tau_{\mathrm{g}}$ is regarded as the group delay, which only induces a temporal shift of the pulse train, while the parameter $\Delta\Phi$ changes the temporal profile. In fact, we can uniquely define the spectral phase $\phi$ as a quadratic function to the angular frequency $\Omega$ such that it passes on the given three points of the spectral phases, $\phi_{11}$, $\phi_{13}$, and $\phi_{15}$, at $\Omega_{11}$, $\Omega_{13}$, and $\Omega_{15}$, respectively. That is,

$$\phi(\Omega) = \frac{\Delta\Phi}{4\omega_{\mathrm{f}}^2}(\Omega - \Omega_{13})^2 + \tau_{\mathrm{g}}(\Omega - \Omega_{13}) + \phi_{13}, \tag{17}$$

thus, the "chirp" at $\Omega_{13}$ should be expressed as, $\ddot{\phi}(\Omega_{13}) = \Delta\Phi/(2\omega_f^2)$. Hence, all we have to do is to determine $\Delta\Phi$, or equivalently $\ddot{\phi}(\Omega_{13})$, if we would like to determine the temporal profile of the intensity of the pulse train formed by three harmonic fields.

We fitted the benchmarks of Eqs. (13) and (15) to the measured autocorrelation traces in Fig. 17, which are obtained by averaging each mode at $\pm 0.5$ eV, at the 24th- and 28th-order modes, respectively, with a constant background, and then found that the correlated part of the measured traces should be $0.283 \pm 0.12$ of the maximum intensity of each mode. The ratios among the intensities of the three harmonic fields were maintained in this procedure. With the assumption that the 26th-order mode includes the correlated part with the same ratio, $I(\Omega_{26}; \tau)$ in Eq. (14) was fitted to the measured trace in Fig. 18 at the 26th-order mode. $|\Delta\Phi|$ and $|\ddot{\phi}(\Omega_{13})|$ were $0.145 \pm 0.013$ radians and $1.3 \pm 0.1 \times 10^{-32}$ s$^2$, respectively.

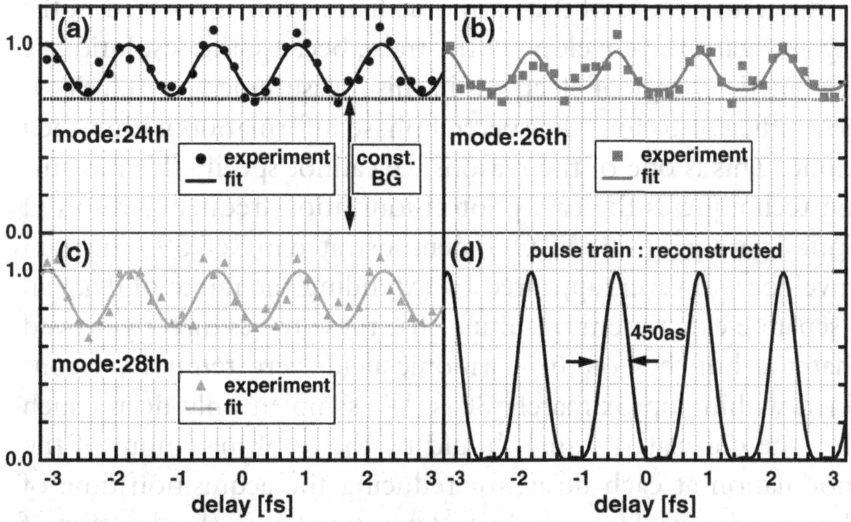


**Fig. 18.** Autocorrelation traces at three individual modes and the pulse train reconstructed. (a) 24th mode, (b) 26th mode, and (c) 28th mode. The temporal profile calculated from the measured chirp and the harmonic distribution is also shown in (d).

Although the signs of $\Delta\Phi$ and $\ddot{\phi}(\Omega_{13})$ cannot be determined due to the nature of the second-order autocorrelation, we assume them to be a positive for the consistency with that in theories[36] and other related works of the experiment. As a result, these quantities agree well with those reported for other measurement techniques[2,36,70] under the condition that the intensity of the fundamental laser field is estimated to be $\sim 10^{14}$ W/cm$^2$. We conclude, by substituting the obtained numeric to Eq. (16), that the pulse duration should be 450 attoseconds, at most, at full width at half maximum (FWHM), which is almost the same as that for the Fourier limit (445 attoseconds), as was shown in Fig. 18(d). This has been the finest temporal intensity profile, to our knowledge, observed by the autocorrelation technique to the date for the next experiment described in Sec. 5.

The photoelectron analysis of nonresonant two-photon-ionization for harmonic electric-field reconstruction (PANTHER)[43] from the mode-resolved autocorrelation traces mentioned above was feasible for determining the characteristic of the train feature of the harmonic field, while it clarified nothing as regard the envelope of the train, because the resolution of the kinetic energy of electrons in this measurement ($\sim$0.5 eV) is not sufficient to find the spectral phase information within each mode. This is one of the reasons we cannot specify this measuring technique as the two-photon-ionization frequency-resolved optical gating[25] (TPI-FROG) demonstrated in Ref. 59. The train envelope, however, appeared by extending the range of delay in a separate experiment, resulting that these traces directly proved that the bunches of the harmonic field were temporally confined within approximately 30 fs. We sampled only delays such that the correlated signal should be the local maximum of the modulation at each delay, for reducing the acquisition time of this experiment. Here, we have determined both the duration of the pulse in the train and the upper limit of the duration of its envelope.

## 5. Interferometric Autocorrelation of APT with Two-Photon Coulomb Explosion

### 5.1. *Similarity of APT with mode-locked laser pulses*

The model of an APT expressed as Eq. (9) is very similar to the model of the pulse train generated from a mode-locked laser oscillator. In order to clarify the importance of an interferometric autocorrelation measurement of an APT, we will briefly explain how the phase relationship between the pulses in the train of the mode-locked laser oscillator can be fixed and measured, according to the research results demonstrated by Hänsch[53] and Hall,[22] who were both awarded the Nobel Prize for physics in 2005.

The positive frequency part of the electric field of the mode-locked pulse train from a laser oscillator, $E^{+}_{\mathrm{ML}}(t)$, can be modeled with an equation of

$$E^{+}_{\mathrm{ML}}(t) = \sum_{m=m_1}^{m_2} F_m \mathrm{e}^{-i\omega_m t - i\varphi_m}, \tag{18}$$

where we define the $m$th frequency mode as

$$\omega_m \equiv 2\pi(m\nu_{\mathrm{rep}} + \delta\nu), \tag{19}$$

with a repetition frequency of the pulse train, $\nu_{\mathrm{rep}}$, which is typically several tens of MHz~100 MHz in a radio frequency range. The frequency offset (FO) $\delta\nu( < \nu_{\mathrm{rep}})$ should be included in each mode frequency because the phase velocity and the group velocity of the laser pulses in the oscillator cavity are generally not the same due to the dispersions of many optical elements inside the cavity. In particular, the additional change of the refractive index with the thermal effect inside the solid state laser material (which is the most common Ti:sapphire), induced by the fluctuation of the pump power of the pumping laser, notably affect the stability of $\delta\nu$. Thus, the FO is generally unstable in a conventional mode-locked laser oscillator, resulting in the interference fringes between different pulses in the train being washed out.

We may be able to stabilize with a closed loop control of the pumping laser field if we can measure $\delta\nu$, but how? The key idea for the detection and the control of $\delta\nu$ was independently proposed by three research groups.[22,53,68] They utilized the two kinds of nonlinear frequency shifts of the mode-locked laser field with different projections to $\delta\nu$ in the shifted frequency in a similar wavelength region, and then those frequency-shifted laser fields can be interfere with an effect due to the FO. For an example of the conventional $f - 2f$ interferometry, the spectral range of the laser field is broadened with a self-phase modulation in a photonic crystal fiber such that the optical frequency-range can be spanned more than 1 octave, then the long wavelength region of the laser is converted to the second harmonic in a nonlinear crystal. The frequency modes of this second harmonic (SH) field of $E^{+}_{\mathrm{ML}}(t)$ include the FO of $2\delta\nu$ because the positive frequency part of the SH field, $E^{+}_{\mathrm{SH}}(t)$, is proportional to $\left\{E^{+}_{\mathrm{ML}}(t)\right\}^2$, while the frequency modes at the shortest region of the broadened spectrum with the SPM, $E^{+}_{\mathrm{SPM}}(t)$, the wavelength of which is similar to that of the second harmonic field, keep the FO $\delta\nu$. The selected frequency modes of $E^{+}_{\mathrm{SH}}(t)$ and $E^{+}_{\mathrm{SPM}}(t)$ in the same wavelength region are proportional to $F'_{m'}e^{-2\pi i(2\nu_{m'}+2\delta\nu)t-2i\varphi_{m'}}$ and $F_m e^{-2\pi i(\nu_m+\delta\nu)t-i\varphi_m}$, respectively under the condition satisfying $2\nu_{m'}\simeq\nu_m$ at certain $m'$ and $m$. Thus, we can observe the beat note on the intensity of the pulse train, $I_{\mathrm{beat}}(t)$, with a frequency of $\delta\nu$, by combining those two fields and selecting the SH wavelength region of the combined field through a narrow band filter, due to the interference term of the two fields as shown in Eq. (20).

$$I_{\mathrm{beat}}(t) \sim |F'_{m'}|^2 + |F_m|^2 + 2F'_{m'}F_m \cos\left(2\pi\delta\nu t + 2\varphi_{m'} - \varphi_m\right). \tag{20}$$

We find that the amplitude of $m'$th mode of the SH field $F'_{m'}$ and that of $m$th mode of the self-phase modulated field $F_m$ should be similar for an appropriate visibility of the interference. In the actual experiment, the bandwidth of the spectral filter is finite, so that all three terms in Eq. (20) include oscillation with a frequency of $\nu_{\mathrm{rep}}$ and with its harmonic frequencies.

In a commercially available FO-stabilized laser oscillator,[14,15] the FO should be locked with a down-counted frequency of the repetition frequency $\nu_{\mathrm{rep}}/m_0$ with a closed feedback loop to the power of the pumping laser, where $m_0$ is arbitrary positive integer. (The DC stabilization satisfying $\delta\nu = 0$ needs more complex system with a frequency shift of the reference frequency.) The phase of the pulse train of this type of FO-stabilized laser oscillator exhibits a time-translation symmetry of

$$E_{\mathrm{ML}}^{+}(t + qT_{\mathrm{rep}}) = \mathrm{e}^{-2\pi iq/m_0} E_{\mathrm{ML}}^{+}(t), \qquad (21)$$

for arbitrary integer $q$, where we define the time interval between the pulses in a train $T_{\mathrm{rep}}$ as $\nu_{\mathrm{rep}}^{-1}$. For an example, we have $E_{\mathrm{ML}}^{+}(t + T_{\mathrm{rep}}) = \mathrm{e}^{-i\pi/2} E_{\mathrm{ML}}^{+}(t)$ by setting $\delta\nu$ to $\nu_{\mathrm{rep}}/4$. It can be recognized from Eq. (21) that the intensity profiles of the pulse, that is proportional to $|E_{\mathrm{ML}}^{+}(t)|^2$, are all the same in each pulse, whilst the electric field itself is reproduced in each $m_0$th pulse. The phase difference between the pulses, however, should be fixed so that we can expect to observe interference between different pulses in the train generated from the FO-stabilized laser oscillator.

We can easily see the similarity of an APT to the FO-stabilized laser pulse train by comparing Eq. (8) with Eq. (18). These two equations should be the same just by replacing $\nu_{\mathrm{rep}} \Rightarrow 2\nu_{\mathrm{f}}$, $\delta\nu \Rightarrow \nu_{\mathrm{f}}$, and $F_m \Rightarrow A_{2n+1}$, respectively. We define the carrier frequency of the fundamental laser field $\nu_{\mathrm{f}} \equiv \omega_{\mathrm{f}}/(2\pi)$ in these substitutions. Thus, the APT formed with synthesized harmonic fields should exhibit the same time-translation symmetry as that contained in the FO-stabilized laser pulse train with a FO of $\nu_{\mathrm{rep}}/2$, of which the symmetry can be expressed as

$$E^{+}(t + qT_{\mathrm{f}}) = E^{+}(t), \tag{22}$$

and,

$$E^{+}\left(t + \frac{q}{2}T_{\mathrm{f}}\right) = \mathrm{e}^{-i\pi}E^{+}(t) = -E^{+}(t). \tag{23}$$

The latter equation suggests that the phase between pulses in an APT should be locked such that the direction of the electric field in the pulse in the APT should be opposite to that of the next pulse with the same intensity profiles. This type of phase-locking with the definite symmetry is a unique feature of an APT that cannot be seen in an isolated attosecond pulse.

### 5.2. *Why do we need interferometric autocorrelation?*

We have found the phase-locking and the symmetry in an APT in Eqs. (22) and (23). How can we directly observe these properties in the APT in an experiment?

One may claim that they are not worth observing because they are intrinsically proven by confirming with an XUV spectrometer that the harmonic spectra correspond to exactly the odd frequencies of the fundamental laser field. We have to notice, however, that the wavelengths of the harmonic spectra can be shifted with an intrinsic chirp of a dipole phase in each harmonic field and/or with a chirp of the fundamental laser field.[27] Because the shift of the wavelength corresponds to the change of the FO, there may be a discrepancy of the time-translation symmetry due to the ambiguity of the absolute wavelengths of the harmonic fields with a limited accuracy for the calibration of the XUV spectrometer. Thus, we would like to emphasize the importance of the direct observation of the phase-locking and the time-translation symmetry in time domain.

Another may propose a sort of experiments based on the TPI-FROG[59] or the PANTHER.[43] In spite of the feasibility for determining the chirp and the intensity profile of the APT, the FROG type measurement cannot reveal the constant phase shift appearing in Eq. (23). The reason can be found in Eq. (11). The FROG or PANTHER essentially measures the absolute square of $\tilde{S}^{+}(\Omega_{2(N+1)};\tau)$, which does not be affected by replacing one of the electric fields in the right-hand side of Eq. (11) to $e^{-i\pi}E^{+}(t)$ or $e^{-i\pi}E^{+}(t-\tau)$. To our regret, we must discard this idea. Because the synthesized optical field of the APT can be expressed as the same as that of the pulse train from the FO-stabilized laser, we should refer to an experiment concerning the pulse train generated from a FO-stabilized mode-locked laser oscillator. Jones *et al.* demonstrated the interferometric "cross"-correlation between the $i$th and $(i+2)$th pulses in a train.[22] They observed a clear correspondence between the FO and the phase of the fringe to the correlation envelope. This result suggested that (i) the phase should be locked between the two pulses such that interference fringes can appear and (ii) the difference in the locked phase between the two pulses should be exactly determined by the FO. Thus, we must perform the experiment based on the interferometric autocorrelation (IAC) of the APT (we prefer the term "auto" to "cross" for the correlation of the APT that is regarded as a united optical field lasting for only a tens or a few of femtosecond duration). We should note that there is nothing but the interference fringes that can reflect a constant phase shift.

### 5.3. *Two-photon Coulomb explosion*

We have a crucial issue for realizing an interferometric autocorrelation measurement by utilizing the electrons via the TP-ATI as a correlated signal. The number of the counted electrons in each two-photon mode in the TP-ATI spectra is rather small so that we should accumulate the data in 4000 shots of the laser pulse in each delay to obtain the sufficient signal-to-noise ratio of the electron spectra shown in Fig. 17. The longer acquisition time requires the higher stability of the laser system. It is not reasonable in practice to increase the resolution of delay to find interferometric fringes with the laser system kept stable in a long time scale more than 1 week or 1 month. Further, a tin filter restricting the bandwidth of the harmonic spectra may disturb the interferometric fringes in principle due to the spatial inhomogeneity of the thickness. Thus, we should find other nonlinear interactions to give us a sufficient correlated signal within a short-term data acquisition, while we do not need an accurate information of chirp, because we have already measured it with the PANTHER.

It may have been expected that a molecule is more likely to be ionized with a two-photon absorption of XUV/soft X-ray light field than an atom because of the dense intermediate states of an electron, due to degrees of freedom for nuclear coordinates, contributing the two-photon transition. In fact, the nonlinear cross section of a hydrogen molecule for the TP-ATI is estimated to be larger than that of a helium atom by approximately one order of magnitude.[20] We have searched for a suitable molecule for two-photon ionization to both singly charged and doubly charged states that may dissociate, resulting in a nitrogen molecule being the most favorable.[51]

We show the potential energy curve of the ground state of an electron in each charge state of molecule as a function of the internuclear distance between two nitrogen atoms in Fig. 19.[51] The evaluated ionization energy of a nitrogen molecule is $\sim 15.6$ eV, thus it can be singly ionized with one-photon absorption of the high-order harmonic field of which order is larger than or equal to 11th, while the direct transition from a neutral molecule to the doubly charged state needs at least the energy corresponding to that of the 28th harmonic field. Hence, the observation of ions originating from a doubly charged state of a nitrogen molecule

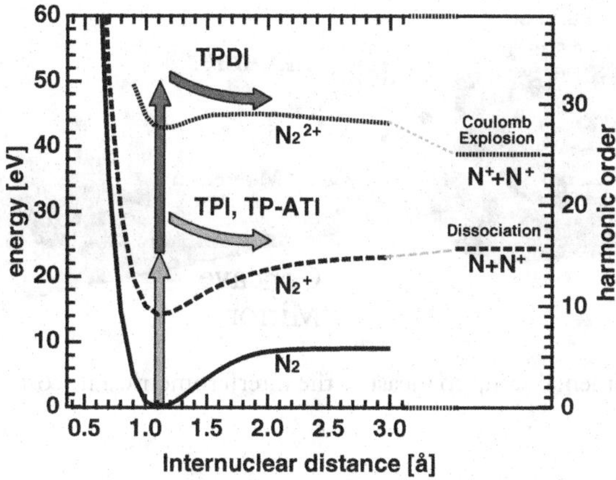


**Fig. 19.** Potential energy curves for $N_2$ (thin solid curve), $N_2^+$(thick dashed curve), and $N_2^{2+}$(thick dotted curve), respectively. Photon energies of two kinds of harmonic field are indicated by arrows with different gray scales. The equivalent orders of the harmonic field of Ti:sapphire laser is shown on the right axis.

should be the evidence of two-photon absorption of the harmonic fields generated from a xenon gas target of which harmonic order does not reach to 29th. In particular, the doubly charged state on a dissociative potential curve, mainly contributed from the repulsion with the Coulomb force, yields two singly charged atomic nitrogen ions having large kinetic energy release, which can be experimentally distinguished from the doubly charged nitrogen molecule in a time-of-flight mass spectrum. Thus, we have concentrated on finding this phenomena, known as the Coulomb explosion of a molecule, with two-photon absorption of the harmonic fields forming the APT (we called this phenomenon as two-photon Coulomb explosion, TP-CE), and then executed the interferometric autocorrelation measurement.

Note that singly charged atomic ion with a dissociation process can also arise from the transition to a excited state of a singly charged nitrogen molecule ion via two-photon ionization (TPI) or two-photon above-threshold ionization, as shown in Fig. 19.

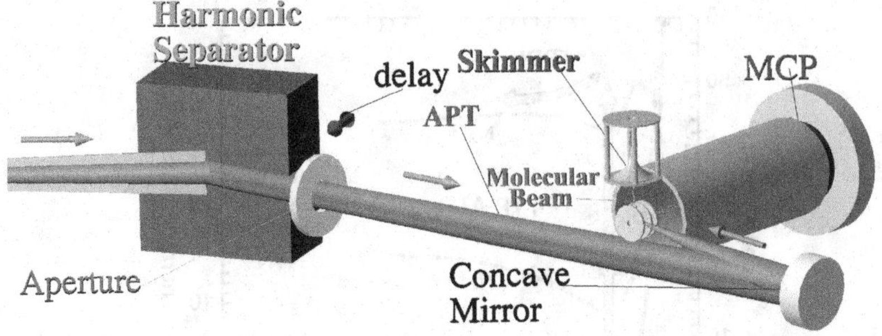


**Fig. 20.** Experimental setup to measure the interferometric autocorrelation trace of an APT.

### 5.4. *Interferometric autocorrelation*

The experimental setup is similar to that in the previous experiment described in Sec. 4, except for the following modifications.[44] The configuration of the setup is shown in Fig. 20.

First, we have replaced the electron spectrometer to an ion mass spectrometer. Second, we remove a tin filter to eliminate the possible degradation of the spatial coherence of the harmonic fields by going through the filter. We reduce, instead, the diameter of an aperture behind the harmonic separator mirrors from 3 mm to 2 mm, and also extend the focal length of a concave mirror made of siliconcarbide from 50 mm to 100 mm, in order to decrease the intensity of the residual fundamental laser field. The intensity of the residual fundamental laser field is suppressed to less than $5 \times 10^{11}$ W/cm$^2$, while 60% of the energy of the harmonic fields passes through the aperture, so that the intensity of the 11th-order harmonic field should have reached $2.8 \times 10^{12}$ W/cm$^2$. We confirmed, in a separate experiment on electron spectroscopy, that the intensity of the fundamental laser field was considerably lower than the detection limit for observing the sidebands of the electron spectra induced with the two-color ATI of an argon atom, whose ionization potential is comparable to that of the nitrogen molecule.

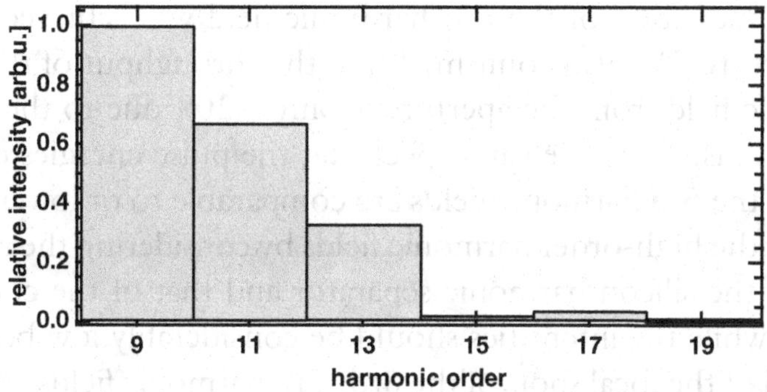


**Fig. 21.** Relative intensities of harmonic fields.

The surface figure of the concave mirror and each harmonic separator mirror are $\lambda/20$ and $\lambda/40$ at 633 nm, and only 1% and 3–5% of the areas on the surfaces of these optics are used to reflect the harmonic fields, respectively. Thus, these mirrors are not expected to distort the wavefront.

One of the benefits of this scheme without using a metal filter is that we can utilize a wide range of harmonic fields for an application. Actually, the measured spectra of the harmonic fields in this experiment are extended from the 9th to the 19th orders, as shown in Fig. 21, whereas we could find only three harmonic fields of 11th, 13th, and 15th orders due to the bandwidth limit of the tin filter that we used in the previous experiment.

There should be a drawback in this scheme. The remaining lower order harmonic fields of 3rd and 5th orders, which can be impeded with a metal filter, may affect the ionization process. Thus, we measured the pulse energy of the 3rd order harmonic field with an energy meter, resulting in an energy of 1.9-$\mu$J before reflection with the silicon harmonic separator mirrors. The relatively low energy of the 3rd harmonic field was due to the low pressure of the xenon gas, which was optimized for the maximum yields of high-order harmonic fields of 11th, 13th, and 15th orders significantly absorbed in the xenon gas medium.

Further reduction of the 3rd harmonic field was achieved with the aperture. We also confirmed that the throughput of the 3rd harmonic field from the aperture is only ~20% due to the large beam radius. Thus, we can expect that the pulse energies of the 3rd and the 5th harmonic fields are comparable to or lower than those of the high-order harmonic fields by considering the reflectivity of the silicon harmonic separator and that of the concave mirror, while the intensities should be considerably low because the radii of the focal spots of the 3rd/5th harmonic fields are 2–3 times larger than that of the high-order field. The relative intensities of the harmonic fields at the focal region in Fig. 21 were determined by observing the photoelectron spectra yielded via the one-photon absorption of an argon atom. We also took into account spectra measured using an XUV spectrometer to estimate the relative intensity of the 9th-order harmonic field, for which the photon energy is lower than the ionization threshold of argon.

The molecular beam was introduced through a skimmer from a pulsed gas valve with a backing pressure of 1–1.5 atm and a pulse duration of 200-$\mu$s. The ions originating from the nitrogen molecules yielded at the focus of the harmonic fields are accelerated using two stages of static electric fields between three plates of electrodes to minimize the aberration of the ion spectrum, or in more specific terms, to satisfy the Wiley–Mclaren condition for the time of flight of ions. The diameters of the bores of the two electrodes are both 1 mm, and the flight length of the ions to an ion detector (microchannnel plate) is 50 cm. The ambient pressure of the time-of-flight chamber is approximately $10^{-6}$ torr. The poor vacuum of the chamber is due to stray xenon gas from the chamber used for the harmonic generation. We measured the time-of-flight mass spectrum using a digital oscilloscope, with a bandwidth of 400 MHz and a sampling rate of 5 GS/s, controlled using the composite software for the moving stage and data acquisition in a desktop computer.

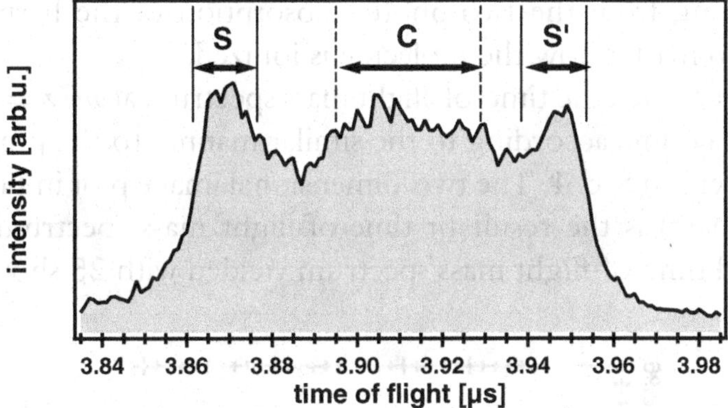


**Fig. 22.** Time-of-flight mass spectrum at $m/z = 14$. Typical spectrum of ions originating from a $N_2$ molecule with two-photon absorption of high-order harmonic fields.

A typical time-of-flight mass spectrum at the mass number $m/z = 14$ is shown in Fig. 22. We can determine that the side peaks, denoted as S and S′ in this figure, are attributed to the singly charged atomic nitrogen ion yielded from the Coulomb explosion of a doubly charged nitrogen molecule, because the kinetic energy release of the ions of the S and S′ peaks is estimated to be 5 eV, which corresponds to the energy difference between the doubly charged nitrogen molecule and its two fragments of the singly charged atomic nitrogen ion.

On the other hand, we can assign the central peak, denoted as C with the smaller kinetic energy release, to singly charged atomic nitrogen ions yielded via the dissociation of singly charged nitrogen molecules due to the small kinetic energy release. Other candidates of ionization process for the C peak are the TP-ATI related to dissociative potential curves of the singly charged nitrogen molecule, and the two-photon double ionization without dissociation. Even though we cannot specify the branching ratio of these ionization processes included in the central peak in Fig. 22, the most important point for measuring the autocorrelation trace of an APT is that the central peak necessarily contains ions arising from the two-photon absorption of the harmonic fields, no matter how the molecule is ionized.

We recorded the time-of-flight mass spectrum at $m/z = 14$ at each delay time according to the similar manner to the previous experiment in Sec. 4. The two-dimensional image plot in Fig. 23 (upper part) is the resultant time-of-flight mass spectrum. We acquired time-of-flight mass spectrum yielded with 25 shots of a

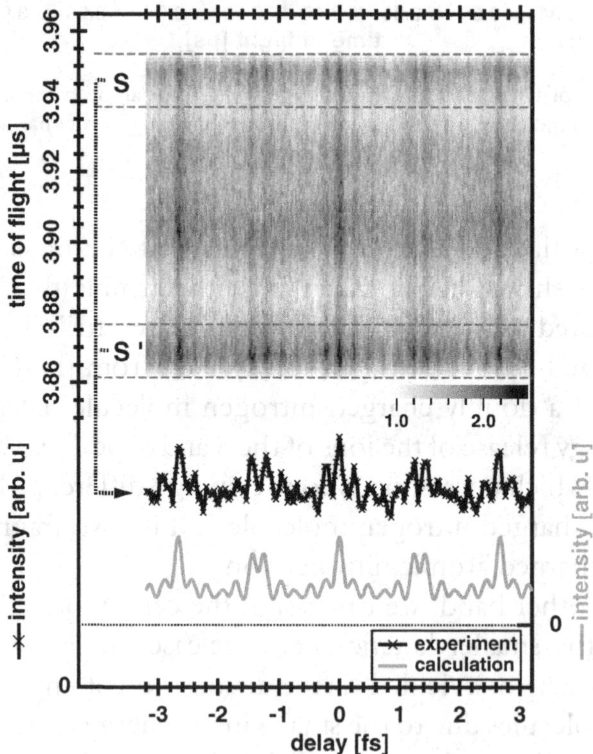


**Fig. 23.** Upper part: Variation of the mass spectrum at $m/z = 14$ by translating the delay between the two replicas of the harmonic fields. Bunches with a period of 1.33 fs and fringes with a much shorter period reveal the interferometric autocorrelation of an APT. The gray scale of the intensity is shown in the inset. Lower part: Interferometric auto correlation trace. We averaged the image within the two regions denoted as S and S′ in the correlation image (crossing markers connected with solid lines). The calculated correlation trace with the group delay dispersion of $1.3 \times 10^{-32}$ [$s^2$] is shown as the gray solid curve.

fundamental laser pulse and scanned the delay every 27 attoseconds from $-3$ to 3 fs. The scanning was repeated 40 times. As a consequence, the time-of-flight mass spectrum at each delay should be the 1000-shot average of the acquired data.

We can see clear bunches with a period of 1.33 fs, which is exactly half the period of the optical frequency of the fundamental laser field ($T_f = 2.67$ fs), throughout the whole spectrum. These bunches are attributed to the autocorrelation of the pulse envelope of the APT, because the separation between adjacent pulses in the APT is $T_f/2$. Note that the appearance of the autocorrelation itself in the time-of-flight mass spectra ensures that there is a nonlinear interaction of the harmonic fields with the ion yields, since a linear interaction cannot provide the correlated signal of the pulse envelope. We can also confirm that the C peak should mainly originate from some kinds of nonlinear process by observation of the autocorrelation at the central region.

In addition to the pulse envelope, we should consider the fine fringes on the autocorrelation trace. To see the details of the fringes, we plot the average trace within the side-peak regions (S, S′), as shown in lower part of Fig. 23. We restrict ourselves to analyzing only the side-peak regions, because we cannot exactly specify how the autocorrelation emerges in the central region (C), as we mentioned before. We can confirm that the fringes on the autocorrelation should originate from the interference between the two replicas of the harmonic fields by carrying out a Fourier transform of the trace in Fig. 23. The absolute square magnitude of the Fourier amplitude of the trace to the frequency, which is normalized using the carrier frequency of the fundamental laser field, is shown in Fig. 24. The peaks at frequencies of $9\nu_f$, $11\nu_f$, and $13\nu_f$ correspond to the interference fringes of 9th-, 11th-, and 13th-order harmonic fields, respectively, whereas the peaks at a frequency of $2\nu_f$ and $4\nu_f$ are evidence that harmonic fields of different frequencies form the envelope of the APT. The absence of expected peaks at frequencies higher than $13\nu_f$ is

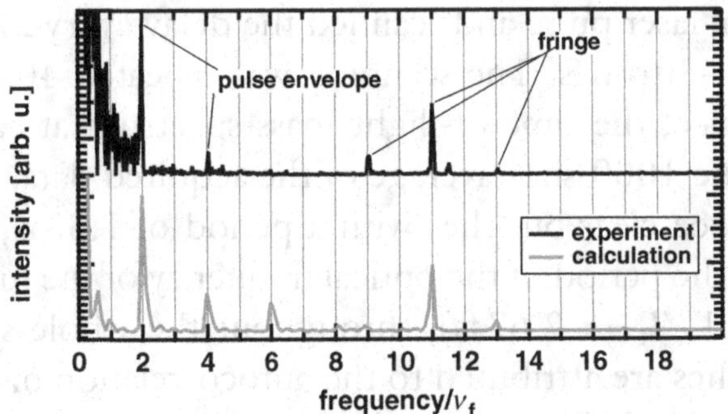


**Fig. 24.** Intensities of the Fourier transforms of the interferometric autocorrelation traces in Fig. 23, for experiment (upper part) and calculation (lower part), respectively. The peaks at frequencies of $2\nu_f$ and $4\nu_f$ are assigned to be the modes of the pulse envelope, while the peaks at $9\nu_f$, $11\nu_f$, and $13\nu_f$ emerge from the interference fringes.

consistent with that of a simulated trace. We did not observe the frequency components of $3\nu_f$ nor $5\nu_f$ in the Fourier transformation of the measured interferometric fringes. This result should be the evidence of the insignificant effect of the 3rd and the 5th order harmonic fields for, at least, the singly charged atomic nitrogen ion yield via TP-CE, which requires higher orders than the 21st for the harmonic fields as a counter part of the 3rd/5th-order harmonic fields.

We should also mention the characteristic of the phase of the fringes to the envelope, other than fringe frequencies, by considering the autocorrelation trace itself. There is a peak of the fringe at the center of the envelope at delays of −2.67, 0, and 2.67 fs, respectively, while a dip in the fringe appears at the center of the envelope at delays of −1.33 and 1.33 fs, respectively. The similarity of the fringe to the time translation of $T_f$ ensures the locked phase between pulses in the APT, at least within the delay range of the observation. Two types of phase in the interference fringes are attributed to the relationship of the phase between adjacent pulses described in Eq. (23). Although the observed fringes originate from the nonlinear interaction, we can determine qualitatively the nature of the fringes by considering only the linear terms of the interference, because the principal frequencies of the fringes correspond not to the sum frequencies but to the carrier frequencies themselves of the harmonic fields, as shown in Fig. 24.

According to Eq. (8), the intensity of the linear summation of $E^+(t)$ and its replica at the next $2q$th pulse with delay can be written as

$$I_q(\tau) \propto \frac{1}{T_f} \int_{-\frac{T_f}{2}}^{\frac{T_f}{2}} dt \left| E^+(t) + E^+(t + qT_f - \tau) \right|^2$$
$$= \sum_{n=n_1}^{n_2} A_{2n+1}^2 \left[1 + \cos\{(2n+1)\,\omega_f \tau\}\right], \qquad (24)$$

whereas that of $E^+(t)$ and its replica at the next $(2q+1)$th pulse is

$$I_{q+\frac{1}{2}}(\tau) \propto \frac{1}{T_f} \int_{-\frac{T_f}{2}}^{\frac{T_f}{2}} dt \left| E^+(t) + E^+\left(t + \left(q + \frac{1}{2}\right) T_f - \tau\right) \right|^2$$
$$= \sum_{n=n_1}^{n_2} A_{2n+1}^2 \left[1 - \cos\{(2n+1)\,\omega_f \tau\}\right]. \qquad (25)$$

The sign exchange in front of the cosine function in these equations reveals the sign exchange or the $\pi$-flipped phase of the fringes for these interferences. Note that at $\tau = 0$, the intensity reaches its maximum in Eq. (24) and in contrast, reaches its minimum in Eq. (25). In fact, these fringes with opposite symmetries should be superposed to the envelope of the autocorrelation trace, so that we observe the peaks and the dips on the top of the pulse envelope in the autocorrelation trace corresponding to these symmetries.

We can explain why the phase of a pulse is opposite to that of the adjacent pulse in the APT from another point of view. According to the 3 step model for the high-order harmonic generation described in Sec. 2.1, the harmonic emission is repeated every half-cycle period of the laser field, whereas the direction of the electric field of the driving laser reverses during a half-cycle period. Because the phase of the emitted harmonic fields is determined by the trajectory of the electron following the direction of the laser field, the phase of the harmonic field slips $\pi$ radians every half-cycle period of the fundamental laser. Thus, our experimental result of the $\pi$-flipped phase on the fringes of the autocorrelation trace is one of the features of the three-step model that should be appeared in time domain.

We estimated the duration of the pulse by comparing the calculated results of the interferometric autocorrelation of the harmonic fields with group-delay dispersion (GDD) and its Fourier transform to the measured results in Figs. 24(a) and 24(b). Please refer to Ref. 61 to find details of this calculation. Setting the group delay dispersion (GDD) to $1.3 \times 10^{-32}$ [$s^2$], which is approximately the same as the measured numeric value with the PANTHER experiment, there is a fairly good agreement between the experimental observation and the calculated result of both autocorrelation and Fourier transform. The increased GDD to $2.0 \times 10^{-32}$ [$s^2$] reduced the height of the envelope with the remaining interference fringe in the autocorrelation so that the intensity at $2\nu_f$ in the Fourier transform is almost equal to that at $11\nu_f$. Therefore, $1.3 \times 10^{-32}$ [$s^2$] is the reasonable GDD in practice, although we might take into account the intrinsic response of a nitrogen molecule to two-photon absorption, as was mentioned in Refs. 46 and 47 for rare-gas atoms, in order to determine GDD in more detail.

The envelope and optical field of the attosecond pulse train, evaluated from the measured intensity ratio of the harmonic fields from the 9th to 19th orders and the estimated GDD, are shown in Fig. 25. The duration of the pulse envelope is 320 attoseconds in full width at half maximum, which corresponds to only an $\sim$1.3 cycle period of the principal carrier frequency of the 11th harmonic field. Note that we owe our success to clearly observing

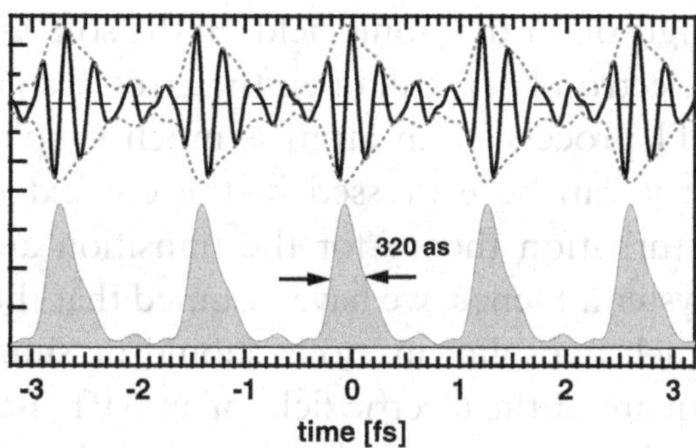


**Fig. 25.** Estimated intensity profile of attosecond pulse train (gary hatched area) and optical field (solid curve).

the symmetry to the short duration of the pulse nearly reaching a single cycle period.

## 6. Summary and Prospects

We have introduced three kinds of nonlinear optics in order to determine directly the temporal characteristic of the high-order harmonic fields of which pulse duration can be shortened to the attosecond regime. The intense light source based on the high-order harmonic generation from the femtosecond laser field is crucial for the experiment of these nonlinear optics.

The TPDI of helium was confirmed by observing the yields of doubly charged helium ion, and then this phenomenon revealed the pulse duration of the 27th harmonic field being sub-10 fs in our research. The most of characteristics of the TPDI is, however, still unknown. The nonlinear crosssection of the TPDI itself is even now under investigation in various theoretical models. Experimental determination of this cross section might be important for testing the theories of TPDI. An electron spectroscopy, which is generally more difficult than an ion spectroscopy as described in Sec. 3, would be needed to determine how the two electrons correlate in the TPDI process. More intense light sources than high-order harmonic fields in the soft X-ray region, such as a free electron laser, will open these research areas.

The TP-ATI process in an atom is much simpler than the TPDI because it can be expressed in the context of the second order perturbation theory for the transition amplitude of one-electron system. Hence, we have assumed that the transition amplitude of each two-photon mode should be simply proportional to the square of the electric field of an APT. Based on this assumption, we have determine the chirp with our PANTHER method by observing the mode-resolved autocorrelation traces, while this experiment is not sufficient to exhibit the full characteristic of the APT. The low resolution of the electron spectrometer cannot resolve the energy within the width of one harmonic spectrum, and the delay range of the correlation does not cover the duration of the train envelope. Therefore, we could not determine how the train continued. Further, the energy splitting of the ground state of an argon ion should be observed in the TP-ATI spectra with the higher resolution. It may disturbs the correct characterization of the APT. We need to improve the resolution for the electron spectroscopy and to find another TP-ATI process in a helium atom without energy splitting for further investigation.

The TP-CE has caused efficient correlated signal for the interferometric auto-correlation measurement of an APT that proves the phase-locking and the time-translation symmetry of the electric field of the APT. We expect that the interferometric fringes should gradually varied or should be degraded around the delay far from the origin because of the intensity-dependent intrinsic phase of the harmonics. This prospect can be confirmed by just shifting the offset of the delay. In addition to the study of an APT, it is interesting to investigate how a nitrogen molecule dissociates by absorbing multiple high-order harmonic fields. The C part of the mass spectrum have revealed the interferometric correlation with different features from that of the S or S′ part.

We cannot, however, find a specific model to describe the experimental result so far, and this is an important issue to be solved in the near future.

## Acknowledgements

Nabekawa and Midorikawa gratefully acknowledge the collaboration with Prof. Kaoru Yamanouch, Univ. of Tokyo. We could not conduct any experiments for the ion and the electron spectroscopy without his supports. We also thank Prof. Akiyoshi Hishikawa, Institute of Molecular Science (IMS) and Prof. Kennosuke Hoshina, Niigata University of Pharmacy and Applied Life Sciences (NUPALS) who were the first collaborators in our laboratory to open the way to the nonlinear optics of ultrafast soft X-ray pulse. The intense harmonic generator was developed by Dr. Eiji J. Takahashi, RIKEN. His struggle to award the world record for the highest pulse energy of high-order harmonic fields played the most important role for achieving our missions. Dr. Hirokazu Hasegawa, IMS spent his all the three years in RIKEN to find the TPDI and the TP-ATI of helium. All the experimental data shown in Sec. 3 are resulted from his painful effort. We thank Dr. Kentaro Furusawa in Canon Inc. for his patience with spending time for the installation and the application of the electron spectrometer in a magnetic bottle although his specialty had been in the research field of fiber lasers. Dr. Toshihiko Shimizu, RIKEN and Dr. Tomoya Okino, Univ. of Tokyo executed both the PANTHER and the interferometric autocorrelation measurements. We cannot forget the impact at the first sight of the interferometric autocorrelation image that they showed us in our laboratory. Nabekawa refers to doctoral theses of Dr. Dai Yoshitomi, National Institute of Advanced Industrial Science and Technology (AIST) and Dr. Tsuneto Kanai, RIKEN for writing Secs. 2.1 and 2.2.

The studies in this chapter were financially supported by MEXT through the advanced optical science project "Extreme Photonics," a Grant-in-Aid for Scientific Research on Priority Areas, for Young Scientists (A), and for Young Scientists (B).

## References

1. P. Antoine, A. L'Huillier and M. Lewenstein, *Phys. Rev. Lett.* 77, 1234–1237 (1996).
2. S. A. Aseyev, Y. Ni, L. J. Frasinski, H. G. Muller and M. J. J. Vrakking, *Phys. Rev. Lett.* **91**, 223902 (2003).
3. P. Balcou, A. S. Dederichs, M. B. Gaarde and A. L'Huillier, *J. Phys. B* **32**, 2973–2989 (1999).
4. P. Balcou, P. Salières, A. L'Huillier and M. Lewenstein, *Phys. Rev. A* **55**, 3204–3210 (1997).
5. A. Bouhal, R. Evans, G. Grillon, A. Mysyrowicz, P. Breger, P. Agostini, R. C. Constantinescu, H. G. Muller and D. von der Linde, *J. Opt. Soc. B* **14**, 950–956 (1997).
6. J. Colgan and M. S. Pindzola, *Phys. Rev. Lett.* **88**, 173002 (2002).
7. E. Constant, D. Garzella, P. Breger, E. Mvel, C. Dorrer, C. L. Blanc, F. Salin and P. Agostini, *Phys. Rev. Lett.* **82**, 1668–1671 (1999).
8. P. B. Corkum, *Phys. Rev. Lett.* **71**, 1994–1997 (1993).
9. M. Drescher, M. Hentschel, R. Kienberger, G. Tempea, C. Spielmann, G. A. Reider, P. B. Corkum and F. Krausz, *Science* **291**, 1923–1927 (2001).
10. L. Feng and H. W. van der Hart, *J. Phys. B* **36**, L1–L7 (2003).
11. M. Ferray, A. L'Huillier, X. F. Li, L. A. Lompre, G. Mainfray and C. Manus, *J. Phys. B* **21**, L31–L35 (1988).
12. D. N. Fittinghoff, P. R. Bolton, B. Chang and K. Kulander, *Phys. Rev. Lett.* **69**, 2642–2645 (1992).
13. E. Foumouo, G. L. Kamta, G. Edah and B. Piraux, *Phys. Rev. A* **74**, 063409 (2006).
14. T. Fuji, J. Rauschenberger, A. Apolonski, V. S. Yakovlev, G. Tempea, T. Udem, C. Gohle, T. W. Hänsch, W. Lehnert, M. Scherer and F. Krausz, *Opt. Lett.* **30**, 332–334 (2005).
15. T. Fuji, J. Rauschenberger, C. Gohle, A. Apolonski, T. Udem, V. S. Yakovlev, G. Tempea, T. W. Hänsch and F. Krausz, *New J. Phys.* 7, 116 (2005).
16. K. Furusawa, T. Okino, T. Shimizu, H. Hasegawa, Y. Nabekawa, K. Yamanouchi and K. Midorikawa, *Appl. Phys. B* **83**, 203–211 (2006).
17. T. E. Glover, R. W. Schoenlein, A. H. Chin and C. V. Shank, *Phys. Rev. Lett.* **76**, 2468–2471 (1996).

18. H. Hasegawa, E. J. Takahashi, Y. Nabekawa, K. L. Ishikawa and K. Midorikawa, *Phys. Rev. A* **71**, 023407 (2005).
19. M. Hentschel, R. Kienberger, C. Spielmann, G. A. Reider, N. Milosevic, T. Brabec, P. Corkum, U. Heinzmann, M. Drescher and F. Krausz, *Nature (London)* **414**, 509–513 (2001).
20. K. Hoshina, A. Hishikawa, K. Kato, T. Sako, K. Yamanouchi, E. J. Takahashi, Y. Nabekawa and K. Midorikawa, *J. Phys. B* **39**, 1–17 (2006).
21. A. Huetz and J. Mazeu, *Phys. Rev. Lett.* **85**, 530–533 (2000).
22. D. J. Jones, S. A. Diddams, J. K. Ranka, A. Stentz, R. S. Windeler, J. L. Hall and S. T. Cundiff, *Science* **288**, 635–639(2000).
23. T. Kanai, X. Zhou, T. Liu, A. Kosuge, T. Sekikawa and S. Watanabe, *Opt. Lett.* **29**, 2929–2931 (2003).
24. T. Kanai, X. Zhou, T. Sekikawa, S. Watanabe and T. Togashi, *Opt. Lett.* **28**, 1484–1486 (2003).
25. D. J. Kane and R. Trebino, *IEEE J. Quantum Electron.* **QE-29**, 580–589 (1993).
26. R. Kienberger, E. Goulielmakis, M. Uiberacker, A. Baltuška, V. Yakovlev, F. Bammer, A. Scrinzi, T. Westerwalbesloh, U. Kleineberg, U. Heinzmann, M. Drescher and F. Krausz, *Nature (London)* **427**, 817–821 (2001).
27. H. T. Kim, D. G. Lee, K.-H. Hong, J.-H. Kim, I. W. Choi and C. H. Nam, *Phys. Rev. A* **67**, 051801R (2003).
28. A. Knapp, A. Kheifets, I. Bray, T. Weber, A. L. Landers, S. Schössler, T. Jahnke, J. Nickles, S. Kammer, O. Jagutzki, L. P. H. Schmidt, T. Osipov, J. Rösch, M. H. Prior, H. Schmidt-Böcking, C. L. Cocke and R. Dörner, *Phys. Rev. Lett.* **89**, 033004 (2002).
29. Y. Kobayashi, T. Sekikawa, Y. Nabekawa and S. Watanabe, *Opt. Lett.* **23**, 64–66 (1998).
30. K. Kondo, A. Sagisaka, T. Tamida, Y. Nabekawa and S. Watanabe, *Phys. Rev. A* **48**, R2531–R2533 (1993).
31. A. Kosuge, T. Sekikawa, X. Zhou, T. Kanai, S. Adachi and S. Watanabe, *Phys. Rev. Lett.* **97**, 263901 (2006).
32. S. Laulan and H. Bachau, *Phys. Rev. A* **68**, 013409 (2003).
33. M. Lewenstein, P. Balcou, M. Y. Ivanov, A. L'Huillier and P. B. Corkum, *Phys. Rev. A* **49**, 2117–2132 (1994).
34. A. L'Huillier and P. Balcou, *Phys. Rev. Lett.* **70**, 774–777 (1993).
35. A. L'Huillier, X. F. Li and L. A. Lompré, *J. Opt. Soc. Am. B* **7**, 527–536 (1990).
36. R. López-Martens, K. Varjú, P. Johnsson, J. Mauritsson, Y. Mairesse, P. Salières, M. B. Gaarde, K. J. Schafer, A. Persson, S. Svanberg, C.-G. Wahlström and A. L'Huillier, *Phys. Rev. Lett.* **94**, 033001 (2005).
37. J. J. Macklin, J. D. Kmetec and C. L. Gordon III, *Phys. Rev. Lett.* **70**, 766–769 (1993).
38. A. McPherson, G. Gibson, H. Jara, U. Johann, T. S. Luk, I. A. McIntyre, K. Boyer and C. K. Rhodes, *J. Opt. Soc. B* **4**, 595–601 (1987).

39. T. Mercouris, C. Haritos and C. A. Nicolaides, *J. Phys. B* **34**, 3789–3811 (2001).
40. N. Miyamoto, M. Kamei, D. Yoshitomi, T. Kanai, T. Sekikawa, T. Nakajima and S. Watanabe, *Phys. Rev. Lett.* **93**, 083903 (2003).
41. Y. Nabekawa, H. Hasegawa, E. J. Takahashi and K. Midorikawa, *Phys. Rev. Lett.* **94**, 043001 (2005).
42. Y. Nabekawa, H. Hasegawa, E. J. Takahashi and K. Midorikawa, *Phys. Rev. Lett.* **97**, 169302 (2006).
43. Y. Nabekawa, T. Shimizu, T. Okino, K. Furusawa, H. Hasegawa, K. Yamanouchi and K. Midorikawa, *Phys. Rev. Lett.* **96**, 083901 (2006).
44. Y. Nabekawa, T. Shimizu, T. Okino, K. Furusawa, H. Hasegawa, K. Yamanouchi and K. Midorikawa, *Phys. Rev. Lett.* **97**, 153904 (2006).
45. T. Nakajima and L. A. A. Nikolopoulos, *Phys. Rev. A* **66**, 041402(R) (2002).
46. T. Nakajima and S. Watanabe, *Phys. Rev. A* **70**, 043412 (2004).
47. L. A. A. Nikolopoulos, E. P. Benis, P. Tzallas, D. Charalambidis, K. Witte and G. D. Tsakiris, *Phys. Rev. Lett.* **94**, 113905 (2005).
48. L. A. A. Nikolopoulos and P. Lambropoulos, *Phys. Rev. Lett.* **73**, 1227–1230 (1994).
49. L. A. A. Nikolopoulos and P. Lambropoulos, *J. Phys. B* **34**, 545–564 (2001).
50. L. A. A. Nikolopoulos and P. Lambropoulos, *J. Phys. B* **40**, 1347–1357 (2007).
51. T. Okino, K. Yamanouchi, T. Shimizu, K. Furusawa, H. Hasegawa, Y. Nabekawa and K. Midorikawa, *Chem. Phys. Lett.* **432**, 68–73 (2006).
52. P. M. Paul, E. S. Toma, P. Breger, G. Mullot, F. Augé, P. Balcou, H. G. Muller and P. Agostini, *Science* **292**, 1689–1692 (2002).
53. J. Reichert, R. Holzwarth, T. Udem and T. Hänsch, *Opt. Commun.* **172**, 59–68 (1999).
54. J. A. R. Samson, Z. X. He, L. Yin and G. N. Haddad, *J. Phys. B* **27**, 887–898 (1994).
55. G. Sansone, E. Benedetti, F. Calegari, C. Vozzi, L. Avaldi, R. Flammini, L. Poletto, P. Villoresi, C. Altucci, R. Velotta, S. Stagira, S. D. Silvestri and M. Nisoli, *Science* **314**, 443–446 (2006).
56. N. Sarukura, K. Hata, T. Adachi, R. Nodomi, M. Watanabe and S. Watanabe, *Phys. Rev. A* **43**, 1669–1672 (1991).
57. J. M. Schins, P. Breger, P. Agostini, R. C. Constantinescu, H. G. Muller, A. Bouhal, G. Grillon, A. Antonetti and A. Mysyrowicz, *J. Opt. Soc. B* **13**, 197–200 (1996).
58. T. Sekikawa, T. Kanai and S. Watanabe, *Phys. Rev. Lett.* **91**, 103902 (2003).
59. T. Sekikawa, T. Katsura, S. Miura, T. Kanai and S. Watanabe, *Phys. Rev. Lett.* **88**, 193902 (2002).
60. T. Sekikawa, A. Kosuge, T. Kanai and S. Watanabe, *Nature (London)* **432**, 605–608 (2004).
61. T. Shimizu, T. Okino, K. Furusawa, H. Hasegawa, Y. Nabekawa, K. Yamanouchi and K. Midorikawa, *Phys. Rev. A* **75**, 044704 (2007).

62. A. A. Sorokin, S. V. Bobashev, K. Tiedtkeand and M. Richter, *J. Phys. B* **39**, L299–L304 (2006).
63. E. Takahashi, Y. Nabekawa and K. Midorikawa, *Opt. Lett.* **27**, 1920–1922 (2002).
64. E. Takahashi, Y. Nabekawa, T. Otsuka, M. Obara and K. Midorikawa, *Phys. Rev. A* **66**, 021802(R) (2002).
65. E. J. Takahashi, H. Hasegawa, Y. Nabekawa and K. Midorikawa, *Opt. Lett.* **29**, 507–509 (2004).
66. E. J. Takahashi, Y. Nabekawa and K. Midorikawa, *Appl. Phys. Lett.* **84**, 4–6 (2004).
67. E. J. Takahashi, Y. Nabekawa, H. M. H. H. A. Suda and K. Midorikawa, *IEEE J. Selct. Top. Quantum Electron.* **10**, 1315–1328 (2004).
68. H. R. Telle, G. Steinmeyer, A. E. Dunlop, J. Stenger, D. H. Sutter and U. Keller, *Appl. Phys. B* **69**, 327–332 (1999).
69. E. S. Toma and H. G. Muller, *J. Phys. B* **35**, 3435–3442 (2002).
70. P. Tzallas, D. Charalambidis, N. A. Papadogiannis, K. Witt and G. D. Tsakiris, *Nature (London)* **426**, 267–271 (2003).
71. H. Wabnitz, L. Bittner, A. R. B. de Castro, R. Döhrmann, P. Gürtler, T. Laarmann, W. Laasch, J. Schulz, A. Swiderski, K. von Haeften, T. Möller, B. Faatz, A. Fateev, J. Feldhaus, C. Gerth, U. Hahn, E. Saldin, E. Schneidmiller, K. Sytchev, K. Tiedtke, R. Treusch and M. Yurkov, *Nature (London)* **420**, 482–485 (2002).
72. B. Walker, B. Sheehy, L. F. DiMauro, P. A. K. J. Shafer, and K. C. Kulander, *Phys. Rev. Lett.* **73**, 1227–1230 (1994).
73. M. Weckenbrock, D. Zeidler, A. Staudte, T. Weber, M. Schöffler, M. Meckel, S. Kammer, M. Smolarski, O. Jagutzki, V. R. Bhardwaj, D. M. Rayner, D. M. Villeneuve, P. B. Corkum and R. Dörner, *Phys. Rev. Lett.* **92**, 213002 (2004).
74. R. Wehlitz, F. Heiser, O. Hemmers, B. Langer, A. Menzel and U. Becker, *Phys. Rev. Lett.* **67**, 3764–3767 (1991).

Chapter 2

# Signatures of Molecular Structure and Dynamics in High-Order Harmonic Generation

Manfred Lein and Ciprian C. Chirilă

*Institut für Physik, Universität Kassel,*
*Heinrich-Plett-Straße 40, 34132 Kassel, Germany*
*Lein@physik.uni-kassel.de*

## 1. Introduction

High-order harmonic generation (HHG)[1,2] is a phenomenon induced by intense laser fields. It refers to the conversion of a large number of laser photons into a single photon of high frequency. The generated frequencies are called high harmonics since they are — for sufficiently long incident laser pulses — integer multiples of the laser frequency. The generated radiation is coherent, and therefore has properties similar to laser light: it emerges as a directed beam and can be temporally compressed into short pulses. The exact mechanism of the multiphoton process that produces harmonics depends on the target of laser irradiation. Harmonics from laser-plasma interaction have received much attention and bear great potential for the future,[3] but these will not be discussed in this review. In the present-day applications, one works mostly with gas-phase harmonics, i.e., radiation emitted from atoms or molecules in a gas jet. Rare-gas atoms are the most frequently used target since they produce harmonics relatively efficiently. Nevertheless the intensity of the generated harmonics is generally small, namely about $10^{-5}$ or less of the incident laser light intensity. In typical experimental setups, laser pulses from Ti:Sapphire oscillators, i.e., with wavelengths around 800 nm are used, and the harmonics range from the ultraviolet into the soft X-ray regime.

The gas-phase harmonics are — at high enough laser intensities — due to a three-step mechanism.[4,5] In the first step, the system (atom or molecule) is ionized by the strong laser field; in the second step, the free electron is accelerated by the oscillating electric field of the laser pulse; and in the third step, the electron returns to the positively charged core and can recombine under emission of a photon. The energy of this photon equals the kinetic energy of the returning electron plus the binding energy of the electron in the ground state of the system. For strong enough laser fields, the excursion length of the electron is much larger than the size of the atom, so that the motion of the continuum electron can be described classically in good approximation. The intuitive three-step model is based on the quasistatic picture of laser-atom interactions: changes of the time-dependent electric field are slow compared to the motion of bound electrons in the atomic ground state. This means that the ionization and recombination can be understood as nearly instantaneous events during which the value of the electric field is constant. The typical laser intensities are between $10^{14}$ and $10^{15}\,\mathrm{W/cm^2}$. The forces that the electric fields of such laser pulses exert on electrons are nearly as high as the forces between the nuclei and the bound electrons. Therefore, the theoretical description must go beyond perturbation theory in the external field. On the other hand, the intensity is still below the regime where the magnetic field of the laser pulses or relativistic motion of the electrons start to play a role. It is therefore safe to employ the dipole approximation, where the magnetic field is neglected and the electron dynamics is described by the nonrelativistic time-dependent Schrödinger equation (TDSE).

The recollision, the final step in the three-step model occurs on a time scale much faster than the optical period of the laser field, which is typically 2–3 fs. Therefore, recollisions give rise to attosecond bursts of harmonic radiation ($1\,\text{as} = 10^{-18}\,\text{s}$.) The generation of attosecond pulses has been one of the main areas of research within laser–matter interactions in recent years.[6–12] The current record in the production of the shortest isolated attosecond pulses is at 130 attoseconds pulse length.[13] With these extremely short pulses it is becoming possible to experimentally observe electronic motion in time, similarly to the observation of chemical reactions with femtosecond pulses. In femtochemistry,[14] the pump–probe scheme is the standard method to observe dynamics: a femtosecond pump pulse starts a time-dependent process, and a femtosecond probe pulse is applied after a delay time to probe the state of the system. Since attosecond pulses are not strong enough yet to allow for attosecond-pump attosecond-probe schemes, one resorts to cross-correlation techniques where an attosecond pulse serves as the pump pulse and an infrared laser pulse probes the time evolution.

Besides using HHG as a light source for further applications, the last years have revealed that harmonics can also be exploited to obtain information about molecular properties.[15–18] Since the recombination step must lead back into the initial state of the molecule in order to give rise to coherent emission, the recombination probability depends strongly on the initial-state electronic wave function. Hence, HHG is very sensitive to the structure of the molecule. Furthermore, it is affected by any deviations of the molecular geometry from the initial geometry caused by the motion of the nuclei in the time between ionization and recombination. We will show in this review how these effects can be used toward the measurement of molecular structure and dynamics. Increasing efforts are presently made to pursue HHG from molecules. In particular, HHG with ensembles of laser-aligned molecules[19,20] plays a central role. For example, alignment techniques opened the possibility to image molecular structure using the molecular-orbital tomography proposed by Itatani *et al.*,[16] which has attracted a tremendous amount of interest. The promise of such schemes is that they combine structure determination with the ultrahigh time resolution offered by femtosecond pulses. Even measurements with sub-femtosecond resolution have been achieved by exploiting the sub-laser-cycle duration of the recollision process[18]: electron wave packets can serve as an attosecond probe and therefore provide an alternative approach to attosecond physics, without the need for attosecond light pulses. This idea was used not only in HHG but also in the earlier work on recollision-induced molecular fragmentation[21,22] where one measures fragment kinetic energies rather than harmonics.

The challenge for theoretical physics is twofold. On one hand, *ab initio* calculations are needed that aim for quantitative agreement with experiment. On the other hand, approximate models are equally important since they give more insight into the mechanisms of HHG, and they provide the basis for schemes of molecular imaging using HHG. The theory of HHG consists of two building blocks: (i) the single-atom/single-molecule response to the laser field and (ii) the propagation of the laser field and the harmonics through the generating medium. We will concentrate on the first part only since we are interested in the signature of molecular properties in HHG.

The purpose of this article is to give a basic and easily accessible introduction into the theory of molecular HHG and to review the current developments in this field. In the following, we will use atomic units, i.e., $m_e = 1$, $e = 1$, $\hbar = 1$, $4\pi\epsilon_0 = 1$ where $m_e$ is the electron mass and $e$ is the elementary charge. This means that the charge of the electron is $q_e = -e = -1$.

## 2. Theory of High-Order Harmonic Generation

### 2.1. *Basic theory*

As explained in Sec. 1, the solution of the time-dependent Schrödinger equation is required to describe theoretically the interaction between molecules and the strong light fields of interest here. The Hamiltonian $H$ consists of the unperturbed part $H_0$ and the laser–molecule interaction $H_{\text{int}}$. In the dipole approximation, the electric field $\mathbf{E}(t)$ of the laser pulse couples to the dipole operator $\mathbf{D}$ of the system so that the interaction is $H_{\text{int}} = -\mathbf{D}\cdot\mathbf{E}(t)$. For fixed nuclei, the electronic dipole operator depends on all the electron coordinates $\mathbf{r}_j$, namely $\mathbf{D} = -\sum_j \mathbf{r}_j$. We begin the discussion of the theory with the single-active-electron (SAE) approximation, where we treat only one of the electrons as interacting with the laser field, while all other electrons remain "frozen," i.e., they merely give rise to an effective, time-independent potential for the active electron. We then have $H_0 = \mathbf{p}^2/2 + V_0(\mathbf{r})$ and $H_{\text{int}} = \mathbf{r}\cdot\mathbf{E}(t)$ with $\mathbf{r}$ being the coordinate of the active electron and $V_0$ the effective potential. The Hamiltonian in the form

$$H(t) = \frac{\mathbf{p}^2}{2} + V_0(\mathbf{r}) + \mathbf{r}\cdot\mathbf{E}(t) \tag{1}$$

is commonly known as the *length-gauge* Hamiltonian because the laser field couples to the "length" $\mathbf{r}$. An alternative treatment is the *velocity gauge*, where the Hamiltonian has the form

$$H(t) = \frac{(\mathbf{p}+\mathbf{A}(t))^2}{2} + V_0(\mathbf{r}) \tag{2}$$

with $\mathbf{A}(t) = -\int_{-\infty}^{t} \mathbf{E}(t')\,dt'$. Here, the laser–molecule interaction $H_{\text{int}} = \mathbf{p}\cdot\mathbf{A}(t) + A(t)^2/2$ is governed by the canonical momentum $\mathbf{p}$, which, for a free electron, can be interpreted as a drift velocity. A third possibility is the treatment in the *acceleration gauge* or *Kramers–Henneberger frame*,[23–27] which is the formulation in the accelerated reference frame of a free electron driven by the laser field. We will not discuss this approach in detail here. Observables have the same value independent of the chosen gauge as long as the TDSE is solved exactly. As soon as approximations are made, different gauges can lead to different results. This happens for example if the calculation is carried out on a too small numerical grid or if one applies approximate schemes such as the strong-field approximation (SFA) that is described later in this chapter.

If the TDSE can be solved, we have the wave function $\Psi(t)$ at all times, and the next task is the calculation of the HHG spectrum. In classical electrodynamics, the spectrum of the dipole radiation emitted from a time-varying charge distribution is the power spectrum of the dipole acceleration, i.e.,

$$S(\omega) \sim |\mathbf{a}(\omega)|^2, \tag{3}$$

where $\mathbf{a}(\omega) = \int dt\, \mathbf{a}(t) \exp(i\omega t)$ is the Fourier transform of the dipole acceleration $\mathbf{a}(t)$. In quantum mechanics, the same equation holds if the time-dependent expectation value of the dipole acceleration is used in place of $\mathbf{a}(t)$, i.e., $\mathbf{a}(t) = \langle \hat{\mathbf{a}}(t) \rangle$.[28–31] The dipole acceleration operator for one electron is $\hat{\mathbf{a}}(t) = \nabla V_0 + \mathbf{E}(t)$. In the last equation, we have taken into account that the electron has negative charge.

The resulting spectrum then contains only the coherent part of radiation, i.e., no spontaneous decay of excited states is included. One may ask what is the appropriate time interval over which the Fourier transformation of the acceleration should be evaluated. In principle, the integration runs over all times. But this is both impossible to do in a numerical calculation and also it would cause problems because an excited state will never decay to the ground state after the laser field has been turned off and therefore oscillations of the acceleration will continue indefinitely. (This is a shortcoming of the TDSE which treats the electromagnetic field classically.) Hence, in a practical calculation one normally works with laser pulses of finite duration $T_\mathrm{p}$ and the Fourier transform is taken only over the duration $T_\mathrm{p}$,

$$\mathbf{a}(\omega) = \int_0^{T_\mathrm{p}} dt\, \mathbf{a}(t) \exp(i\omega t). \tag{4}$$

Nevertheless, one may sometimes choose a somewhat longer integration interval in order to obtain a more accurate harmonic spectrum. Notice that alternatively the spectrum can be calculated by first evaluating the time-dependent dipole moment and then taking the double derivative to obtain the dipole acceleration. This choice plays an important role in the context on the strong-field approximation, see Sec. 2.3. However, when harmonic spectra are calculated from the numerical solution of the TDSE, the method of choice is to evaluate the time-dependent dipole acceleration $\mathbf{a}(t)$ directly as the expectation value of the acceleration operator. It is less sensitive than the dipole moment to electron density far away from the nucleus and is thus less affected by the use of absorbing boundary conditions. Therefore, the acceleration form usually produces a lower background noise level in the HHG spectra.

In the photon picture, HHG is understood as the absorption of a large number of photons from the incident laser field followed by emission of the absorbed energy in the form of one high-frequency photon, see Fig. 1. If the laser pulse has a small

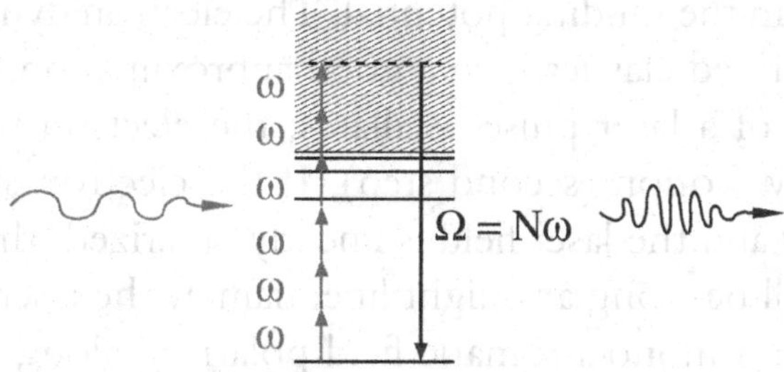


**Fig. 1.** Illustration of high-order harmonic generation as absorption and emission of photons.

bandwidth, i.e., the laser photon frequency has a sharp value, the harmonic emission spectrum will exhibit discrete frequencies, which are approximately integer multiples of the laser frequency.

### 2.2. *Three-step model*

A physical picture that is extremely useful to understand HHG is the semiclassical three-step model.[4,5] It rests on the assumption that the laser–molecule interaction is approximately quasistatic, i.e., it is meaningful to describe the interaction in terms of an instantaneous electric field. At time $t$ during the action of a laser pulse, the field-free binding potential $V_0$ for the active electron is distorted by the presence of the instantaneous electric field $\mathbf{E}(t)$. The total potential is

$$V(\mathbf{r}) = V_0(\mathbf{r}) + \mathbf{E}(t) \cdot \mathbf{r}. \tag{5}$$

In the quasistatic picture, the change of the electric field is slow compared to the bound-state electronic motion so that the electron wave function adjusts to the modified potential and the electron has time to tunnel through the potential barrier into the exterior region. The tunneling ionization process is shown schematically in Fig. 2. After ionization at a time $t_0$ (first step), the electron is strongly accelerated by the laser field with little influence from the binding potential. The electron dynamics can then be described classically in a good approximation. Since the electric field of a laser pulse oscillates, the electron will follow an oscillatory motion (second step). If the electron starts with zero velocity and the laser field is linearly polarized, the classical trajectory will be along a straight line, namely the laser polarization axis. For a monochromatic field polarized along the $z$-axis with electric field strength $E(t) = E_0 \sin(\omega t)$, one can immediately solve Newton's equation of motion, and one finds that the

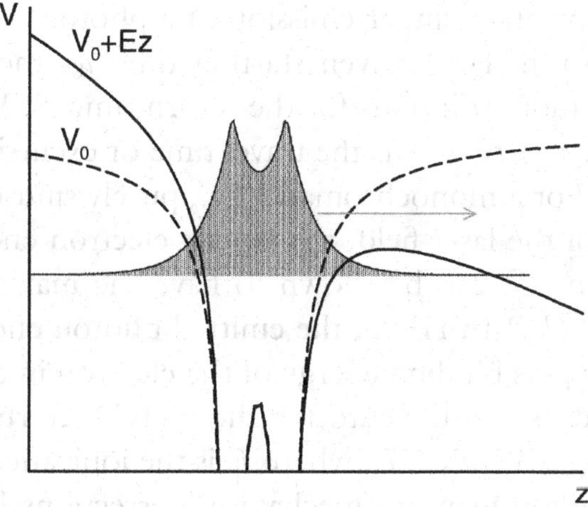


**Fig. 2.** Tunneling ionization in a diatomic molecule under the influence of an instantaneous electric field $E$ directed along the negative $z$-axis.

electron velocity $\dot{z}(t)$ is the sum of a sinusoidally oscillating term and a drift velocity,

$$\dot{z}(t) = \alpha\omega\cos(\omega t) - \alpha\omega\cos(\omega t_0). \tag{6}$$

Here,

$$\alpha = \frac{E_0}{\omega^2} \tag{7}$$

is the classical oscillation amplitude. The oscillation gives rise to a higher mean kinetic energy than from the drift velocity $v_D = -\alpha\omega\cos(\omega t_0)$ alone. This additional kinetic energy due to the presence of the laser field is known as the ponderomotive potential

$$U_p = \frac{E_0^2}{4\omega^2}. \tag{8}$$

If the electron starts at a suitable phase of the laser field, the trajectory can lead back to the core so that an electron–ion collision takes place (third step). With a small probability, the recollision causes recombination under emission of a photon — the high-harmonic photon. For a given starting time $t_0$, there may be zero, one, or more solutions for the return time $t_1$. We refer to the difference $\tau = t_1 - t_0$ as the travel time or excursion time of the electron. For a monochromatic, i.e., purely sinusoidal time-dependence of the laser field, the kinetic electron energy at the time of return, $E_r$, can be shown to have the maximum value $E_{r,max} \approx 3.17\ U_p$.[5] In HHG, the emitted photon energy equals return energy plus binding energy of the electron in the ground state. This means that the cutoff in the HHG spectrum is at the photon energy $3.17\ U_p + I_p$, where $I_p$ is the ionization potential of the molecule. Quantum mechanical corrections lead to the slightly modified cutoff $3.17\ U_p + 1.32\ I_p$.[32]

A given return energy can be realized by not only one but at least two different electron trajectories, known as the short and long trajectories. Most HHG experiments are set up such that — taking advantage of the phase matching conditions for the generation of harmonics in a focused beam — only the short trajectories contribute significantly to the HHG spectrum.[33–35] For these short trajectories, one can fit the relation between the return energy and the travel time obtained from the classical model as

$$\omega\tau = 0.786[f(E_r/U_p)]^{1.207} + 3.304[f(E_r/U_p)]^{0.492}, \quad (9)$$

where $f(x) = \arccos(1 - x/1.5866)/\pi$. The fact that different return energies correspond to different return times $t_1$ makes the harmonic frequency dependent upon the emission time, i.e., the harmonic radiation is *chirped* on a sub-femtosecond time scale.[7]

From the three-step model it is apparent that harmonics are usually strongly suppressed when elliptical or circular laser polarization is used, because electrons starting with zero velocity do not return to the parent ion. We will see below, however, that an anomalous ellipticity dependence of harmonics may result from the initial electron momentum distribution.

### 2.3. *The strong-field approximation*

To arrive at a model that is capable of quantitatively predicting HHG spectra, we formulate the quantum mechanical version of the three-step model, which is known as the Lewenstein model or the strong-field approximation.[32] In the most widely used version of this theory, one first derives an expression for the time-dependent dipole moment $\mathbf{D}(t)$ due to the response of a single atom/molecule. Its dipole acceleration is then obtained as

$$\mathbf{a}(t) = \ddot{\mathbf{D}}(t), \tag{10}$$

and the HHG spectrum is calculated as described in Sec. 2.1. We restrict ourselves to the single-molecule response and do not investigate propagation effects. Furthermore, we begin with the length-gauge formulation based on the Hamiltonian (1) and with the assumption of fixed nuclei. In the strong-field approximation, one assumes that (i) the electron is unaffected by the laser field until the time $t'$ of ionization and (ii) afterwards the electron moves in the laser field only, unaffected by the binding potential. We also neglect the depletion of the bound state (which could be taken into account in principle). For a linearly polarized laser pulse with arbitrary electric field $E(t)$ such that $E(t < 0) = 0$, this leads to the dipole moment (without terms corresponding to continuum–continuum transitions)[32]:

$$\begin{aligned}\mathbf{D}(t) = -i \int_0^t \mathrm{d}t' \int \mathrm{d}^3 p \, \mathbf{d}_\mathrm{r}^*(\mathbf{p} + \mathbf{A}(t)) d_\mathrm{i}(\mathbf{p} + \mathbf{A}(t'), t') \\ \times \exp(-\, iS(\mathbf{p}, t', t)) + \text{c.c.}\end{aligned} \tag{11}$$

where $S(\mathbf{p}, t', t) = \int_{t'}^{t} \mathrm{d}t''[(\mathbf{p}+\mathbf{A}(t''))^2/2 + I_\mathrm{p}]$ is the semiclassical action, $I_\mathrm{p}$ is the ionization potential, $\mathbf{A}(t) = -\int_{-\infty}^{t} \mathbf{E}(t')dt'$, and the ionization and recombination matrix elements are given by

$$d_i(\mathbf{p}, t) = \langle \psi_\mathrm{PW}(\mathbf{p}) | E(t) z | \psi_0 \rangle, \tag{12}$$

$$\mathbf{d}_\mathrm{r}(\mathbf{p}) = \langle \psi_\mathrm{PW}(\mathbf{p}) | - \mathbf{r} | \psi_0 \rangle. \tag{13}$$

Here, $\psi_0$ is the field-free ground state, and $\psi_{\rm PW}(\mathbf{p})$ denotes a plane-wave state with momentum $\mathbf{p}$, normalized in the momentum scale, i.e., $\langle\psi_{\rm PW}(\mathbf{p})|\psi_{\rm PW}(\mathbf{p}')\rangle = \delta(\mathbf{p}-\mathbf{p}')$. The appearance of plane waves is the essential point of the SFA. It facilitates partially analytical computation, but is the main source of errors at the same time.

The SFA expression (11) for the electronic dipole moment can be obtained in different ways: one way is to follow the original approach of Lewenstein *et al.*,[32] where an ansatz is made for the wave function. This ansatz, together with the aforementioned assumptions about the physical process leads to Eq. (11). Another way, more straightforward, is based on the integral equation for the evolution operator:

$$\hat{U}(t,t') = \hat{U}_0(t,t') - i\int_{t'}^{t} dt''\, \hat{U}(t,t'')E(t'')z\hat{U}_0(t'',t'), \quad (14)$$

where $\hat{U}_0$ stands for the evolution operator associated with the field-free Hamiltonian. The integral equation can be approximated by replacing on the right-hand side the full evolution operator $\hat{U}$ by the Volkov propagator $\hat{U}_{\rm V}$. The latter describes the evolution of an unbound electron under the influence of the electric field of the laser only. The replacement assumes that the main influence on the electron dynamics comes from the laser field and less from the binding Coulomb potential, which does not play any role between the ionization time $t''$ and the recombination time $t$ in (14). As a consequence, the low-energy part of the harmonic spectrum is poorly represented. One reason is that in this energy interval the bound–bound dynamics plays a significant role. As a general remark, low-energy electrons (responsible for the emission of low-energy harmonics) "feel" more than high-energy electrons the influence of the binding potential. For this reason, the SFA does not reproduce satisfactorily the above-threshold ionization spectrum close to zero kinetic energy of the emitted electron, but it describes well the high-energy electrons.

With this in mind, the SFA expression for the evolution operator reads

$$\hat{U}^{\mathrm{SFA}}(t,t') = \hat{U}_0(t,t') - i\int_{t'}^{t} dt''\, \hat{U}_{\mathrm{V}}(t,t'')E(t'')z\hat{U}_0(t'',t'). \tag{15}$$

Once the evolution operator is approximated as in Eq. (15), the electronic dipole moment can be easily calculated by inserting the approximate wave function $\psi(t) = \hat{U}^{\mathrm{SFA}}(t,0)\psi_0$ in the expression for the electron dipole moment $\mathbf{D}(t) = \langle\psi(t)| - \mathbf{r}|\psi(t)\rangle$. The continuum–continuum transitions (i.e., the terms containing two integral terms from the right-hand side of Eq. (15)) are ignored, since they do not contribute significantly to the harmonic spectrum. One intermediate step to obtaining the result (11) is to replace the Volkov propagator by its spectral decomposition:

$$\hat{U}_{\mathrm{V}}(t,t') = \int d^3p|\psi_{\mathrm{V}}(\mathbf{p},t)\rangle\langle\psi_{\mathrm{V}}(\mathbf{p},t')|, \tag{16}$$

with $\psi_{\mathrm{V}}(\mathbf{p},t)$ the Volkov solution[36] in the length gauge of an electron with canonical momentum $\mathbf{p}$:

$$|\psi_{\mathrm{V}}(\mathbf{p},t)\rangle = |\psi_{\mathrm{PW}}(\mathbf{p}+\mathbf{A}(t))\rangle \exp\left(-i\int^{t} dt''\frac{(\mathbf{p}+\mathbf{A}(t''))^2}{2}\right). \tag{17}$$

The Volkov solutions describe a free electron moving in the electric field of the laser only. The resulting integral over momenta in Eq. (11) can be approximated as described in the following.

In practice, saddle-point approximations are additionally applied to the integral in Eq. (11).[32,37] The most straightforward approach is to regard the matrix elements as slowly varying functions of the momentum $\mathbf{p}$ while the phase factor $\exp(-iS)$ oscillates rapidly as a function of $\mathbf{p}$. One can then find the saddle-point momentum $\mathbf{p}_s$ where the phase is stationary, i.e., $\nabla_{\mathbf{p}} S(\mathbf{p}, t', t)|_{\mathbf{p}=\mathbf{p}_s} = 0$,

$$\mathbf{p}_s(t', t) = -\int_{t'}^{t} \mathbf{A}(t'')dt''/(t - t'). \tag{18}$$

The saddle-point approximation assumes that the integral is dominated by the integrand around $\mathbf{p}_s$. Thus the matrix elements are evaluated at $\mathbf{p}_s$, and the action is replaced by a second-order Taylor expansion in $\mathbf{p}$ around $\mathbf{p}_s$. Carrying out the integration over momenta leads to[32,37]

$$\begin{aligned} \mathbf{D}(t) = -i \int_0^t \mathrm{d}t' & \left[\frac{2\pi}{\epsilon + i(t - t')}\right]^{3/2} \\ & \times \mathbf{d}_r^*(\mathbf{p}_s(t', t) + \mathbf{A}(t)) d_i(\mathbf{p}_s(t', t) + \mathbf{A}(t'), t') \\ & \times \exp\left(-iS(\mathbf{p}_s(t', t), t', t)\right) + \text{c.c.} \end{aligned} \tag{19}$$

So far, we have introduced the standard form of the Lewenstein model, but alternatively one may calculate the dipole acceleration by directly evaluating the expectation value of the dipole acceleration operator $\nabla V + \mathbf{E}(t)$, without taking the double time derivative in Eq. (10). Or one may evaluate first the expectation value of the dipole momentum operator $i\nabla$ and subsequently take one time derivative to arrive at the dipole acceleration. In Ref. 38, it was shown for $H_2^+$ that the third possibility, the *momentum or velocity form*, gives the best results in the sense that the shape of the resulting HHG spectra is closest to the spectra from the numerically exact solution of the TDSE. In the velocity form, one replaces in Eq. (19) the matrix element $\mathbf{d}_r$ by the dipole-velocity matrix element

$$\mathbf{v}_r(\mathbf{p}) = \langle \psi_{PW}(\mathbf{p}) | \mathrm{i}\nabla | \psi_0 \rangle. \tag{20}$$

After this replacement, Eq. (19) yields not the expectation value of the dipole moment, but of the dipole velocity.

So far, we have not discussed the question of length gauge versus velocity gauge. Although there is not a unique opinion in the literature about what gauge is most appropriate, previous work suggests that length gauge is preferable for atoms and small molecules.[39,81] This seems to happen because in the length gauge the field-dressed ground state is similar to the field-free state used in the SFA. However, this statement does not hold for molecules at very large internuclear separations $R$, as was shown in Ref. 37. In this case, the length gauge leads to an unphysical increase of the cutoff frequency and the harmonic intensities as a function of $R$. Using velocity gauge in combination with a saddle-point approximation adapted to the presence of two centers leads instead to physically reasonable results. In velocity gauge, Eq. (11) is replaced by

$$\mathbf{D}(t) = -i\int_0^t \mathrm{d}t' \int \mathrm{d}^3p\, \mathbf{d}_{\mathrm{r}}^*(\mathbf{p})v_{\mathrm{i}}(\mathbf{p},t')\exp\left(-iS(\mathbf{p},t',t)\right) + \text{c.c.} \tag{21}$$

with the velocity-gauge ionization matrix element

$$v_i(\mathbf{p},t) = \langle \psi_{\mathrm{PW}}(\mathbf{p})| - i\nabla\cdot\mathbf{A}(t) + A^2(t)/2|\psi_0\rangle. \tag{22}$$

The difference with respect to the length gauge is twofold: (i) the operator in the matrix element (22) is the one from the velocity-gauge Hamiltonian (2), and (ii) the momenta at which the ionization and recombination matrix elements in Eq. (21) are evaluated are the canonical "drift" momenta $\mathbf{p}$ instead of the instantaneous momenta $\mathbf{p}+\mathbf{A}(t)$. It is important not to confuse the velocity gauge of the Hamiltonian with the velocity form of the recombination matrix element, Eq. (20). Both can be, but need not be, used in combination with each other.

The internuclear distances at which the unphysical behavior of the length gauge becomes relevant are of the order of $\pi\alpha$ where $\alpha$ is the classical oscillation amplitude of the electron in the laser field. These distances are much larger than typical equilibrium bondlengths. In the following, we will concentrate on the case of small molecules and restrict ourselves to the length-gauge formulation.

## 2.4. *Odd and even harmonics*

For a monochromatic field or sufficiently long pulses, it follows from the photon picture that the harmonic frequencies are integer multiples of the laser frequency. In a typical experiment, one observes only the odd harmonics. This is the consequence of the inversion symmetry of the target gas. In simple terms, all half optical cycles of the driving laser field generate harmonics with the same intensities but with phase differences of $\pi$ relative to each other. The field of the XUV radiation obeys the symmetry

$$\mathbf{E}_{\mathrm{XUV}}(t + T/2) = -\mathbf{E}_{\mathrm{XUV}}(t), \tag{23}$$

where $T$ is the duration of the laser optical cycle. The Fourier spectrum of such a signal has only odd frequency components, i.e., the peaks are separated by twice the laser photon frequency. Equation (23) is a special case of a dynamical symmetry in HHG. Reference 41 discusses the relation between dynamical symmetries and selection rules in HHG.

The existence of inversion symmetry is obvious for atoms which are spherically symmetric. It is also obvious for homonuclear diatomic molecules. In general, however, a molecule with fixed nuclear positions does not possess inversion symmetry. Nonetheless, in an experiment with randomly oriented molecules, one can consider the whole ensemble of molecules as being again inversion symmetric. The same statement remains true in experiments with laser-aligned molecules because alignment does not fix the orientation of the molecular axes. (Orientation refers to knowing also which atom is on which side.) Therefore, in practice one usually observes only odd harmonics when using long laser pulses.

Even harmonics could be measured with oriented heteronuclear molecules. A particularly interesting case of symmetry breaking occurs in an oriented HD molecule. For fixed nuclei, the electron dynamics is completely inversion symmetric. By including the coupling to the vibrational degree of freedom, however, the molecule becomes asymmetric, giving rise to substantial even-harmonic generation.[42] This is an example of non-Born–Oppenheimer strong-field dynamics. Figure 3 compares HHG in one-dimensional $H_2$ and oriented one-dimensional HD. The spectra result from the numerical solution of the TDSE with three degrees of freedom: two electron coordinates and the

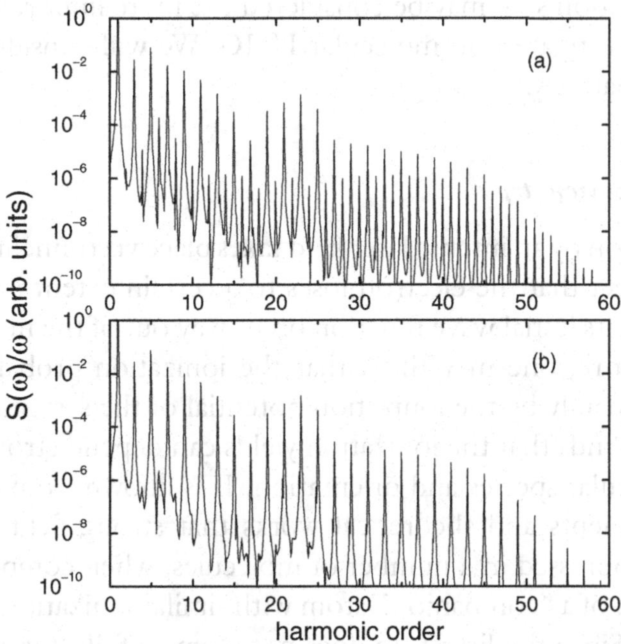


**Fig. 3.** (a) Harmonic spectrum generated from the 1D HD molecule driven by a laser field with peak intensity $10^{14}$ W/cm$^2$ and wavelength 770 nm. The plotted quantity is proportional to the number of emitted photons per frequency interval. (b) The same calculation for the 1D $H_2$ molecule. From Ref. 42, copyright 2001 by the American Physical Society.

internuclear distance. The spectra clearly show the absence of even harmonics in $H_2$ and their presence in oriented HD.

## 3. Influence of Molecular Structure on HHG

Equation (19) gives insight into the influence of molecular structure on HHG. The ground-state molecular orbital appears in the ionization and recombination matrix elements, $d_i$ (or $v_i$) and $\mathbf{d}_r$ (or $\mathbf{v}_r$). We will see that the effect of the two dependences is quite different: the ionization part influences mainly the overall efficiency of HHG and its dependence on the orientation of the molecular axis, while the recombination part additionally determines directly the shape of the HHG spectrum. Therefore, the recombination step may be considered the more important subject of investigation in molecular HHG. We will consider both points separately.

### 3.1. *Ionization step*

If ionization by an intense laser field takes place via tunneling, one may suspect that the electron loses to a certain extent its memory about its initial wave function on its way out of the molecule. Furthermore, one may think that the ionization probability is dictated mainly by the ionization potential of the system. However, one finds that the ionization yields can depend strongly on the molecular species and orientation. It is known from a series of experiments and theoretical works that strong-field ionization is suppressed in a number of molecules, when compared to ionization of a "companion" atom with similar ionization potential.[43–55] Since the literature does not seem to fully agree about the mechanisms responsible for these deviations, we will not further discuss ionization suppression in detail. We note, however, that the ionization step crucially influences the overall efficiency of HHG. We are predominantly interested in the momentum distributions of electron wavepackets after ionization, since they will influence the shape of the HHG spectra. Often, this distribution is found to be rather insensitive to the molecular structure, e.g., in the comparison of aligned $N_2$ molecules at different angles.[56] On the other hand, there is an important exception in the case of molecular orbitals with mirror antisymmetry. If the electric field of the laser pulse points parallel to a nodal plane, then the antisymmetry is conserved, and there are no electrons moving strictly parallel to the field. Rather, the electron always exhibits a lateral drift motion, which makes the usual three-step model inapplicable (since it assumes zero initial velocity). The lateral drift strongly suppresses recollisions, and the efficiency of HHG is reduced. In this exceptional case the use of elliptical polarization can enhance HHG by compensating the drift. We illustrate this behavior[57] using the example of the first excited state of the $H_2^+$ molecular ion, which possesses a nodal plane perpendicular to the molecular axis. Instead of the three-dimensional system, we consider a 2D model system with the binding potential

$$V_0(x,y) = -\sum_{k=1,2} \frac{1}{\sqrt{(x-x_k)^2+(y-y_k)^2+0.5}}. \tag{24}$$

The soft-core potential with the softening parameter 0.5 avoids the numerical difficulties of a Coulomb singularity. It is also physically more applicable than a bare $1/r$ Coulomb potential in 2D, because it mimics the fact that in 3D, the electron has more available "space" to bypass the nucleus. The soft-core potential can be viewed as an average of the 3D potential over the third dimension.[58] The TDSE for one electron in the potential (24) plus the interaction with a 780 nm laser pulse with intensity $4\times10^{14}\,\mathrm{W/cm^2}$ is solved numerically by means of the split–operator method.[59] We consider the special case that the molecule is aligned perpendicular to the major axis of the elliptical laser polarization since this means that the electric field points

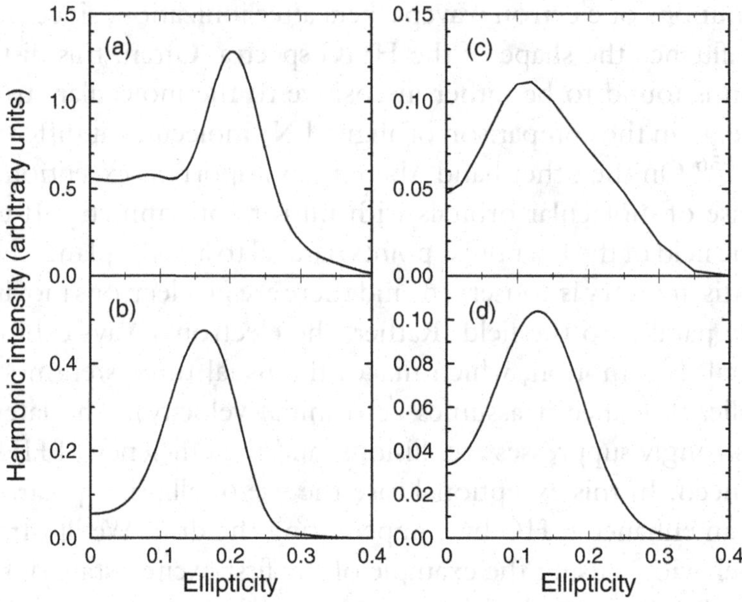


**Fig. 4.** Ellipticity dependence of the harmonic intensities for the antisymmetric first excited state of 2D $H_2^+$, oriented perpendicular to the laser field. Laser pulses with 780 nm wavelength and intensity $4 \times 10^{14}$ W/cm$^2$ are used. (a) 31st, (b) 41st, (c) 51st, (d) 61st harmonic. Data from Ref. 57.

along the nodal plane in the limit of linear polarization. The intensities of the harmonic orders 31, 41, 51, and 61 are shown in Fig. 4 as a function of laser ellipticity. The maximum at non-zero ellipticity is a clear signature of the orbital antisymmetry. It is apparent that the saddle-point version of the strong-field approximation, Eq. (19) cannot explain the non-zero harmonic intensity for linear polarization, since the saddle-point momentum is parallel to the polarization axis and therefore the ionization and recombination matrix elements vanish due to the orbital antisymmetry. This can be interpreted such that the lowest-order saddle-point approximation is not sufficient to replace the integration over momenta in the SFA expression.

Nodal planes induce a strong dependence of the efficiency of HHG on the alignment of the molecule. We note that the antisymmetry leads also to strong suppression of the total ionization probability when the electric field points along a nodal plane so that the orientation dependence of the ionization probability gives a qualitative image of the initial electron orbital.[60]

In the example above there are two physical ingredients, namely a nodal plane and the use of ellipticity, that are both detrimental to the HHG efficiency if they are taken by themselves. In the combination of both, however, they can compensate each other. In the same spirit, it has recently been proposed to compensate the lateral drift from the nodal plane by a non-dipole effect.[61,62] It is well known that the effect of the magnetic field of a laser pulse is to push the freed electron into the laser propagation direction and thereby to reduce the recollision probability.[63,64] For suitable alignment of the molecule, the magnetically induced drift partially counteracts the drift due to the orbital structure. To be precise, the compensation is effective when the laser propagation axis is perpendicular to the nodal plane.

### 3.2. *Recombination step*

In the framework of the SFA, the recombination step affects HHG through the recombination matrix element. The essential difference with respect to the ionization step, however, is that the returning electron momenta are significantly higher than the initial momenta. The de Broglie wavelength of the electron can be comparable to the internuclear distance. Consequently, one may expect interference effects in HHG from a diatomic molecule, in analogy to Young's double slit interference. We demonstrate this behavior by investigating once more the 2D model of $H_2^+$ that we introduced in the previous section. The initial state is now taken to be the symmetric ground state. The resulting spectra for 2D $H_2^+$ aligned at various different angles with respect to the polarization axis of a linearly polarized 780 nm laser pulse are shown in Fig. 5. Two different laser intensities are compared.

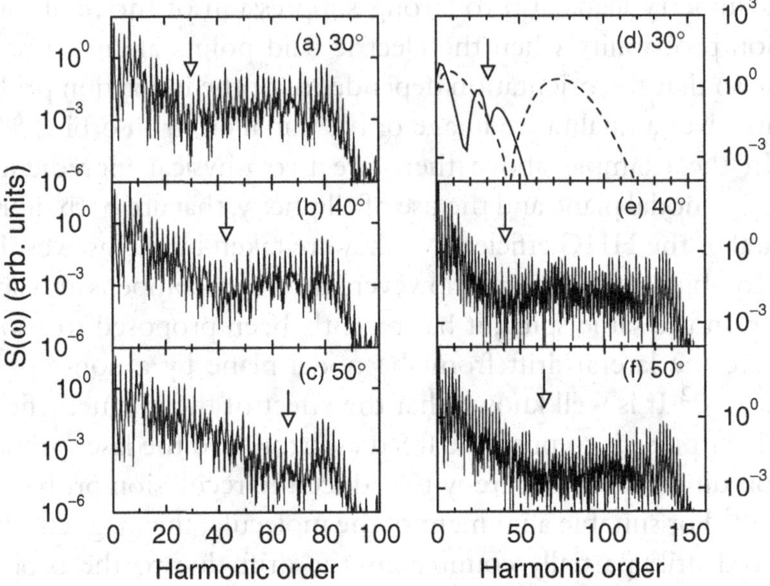


**Fig. 5.** Spectra of harmonics polarized parallel to the laser polarization direction for 2D $H_2^+$ in 780 nm laser pulses, aligned at various angles as indicated. Left panels: laser pulses with intensity $5\times10^{14}$ W/cm$^2$. Panel (d): field-free simulation for recolliding wave packets with energies corresponding to the 31st harmonic (solid line) and 75th harmonic (dashed line). Panels (e, f): laser pulses with intensity $1\times10^{15}$ W/cm$^2$. From Ref. 15, copyright 2002 by the American Physical Society.

Furthermore, panel (d) of the figure shows the result of a laser-field-free simulation where the initial state is the superposition of the ground state with an incoming electron wave packet, the momentum of which is chosen such that it corresponds to the 31st or 75th harmonic, respectively. In all cases, we observe a clear suppression of harmonic emission around a harmonic frequency that depends on the orientation angle but is independent of the laser parameters. The same observation can be made in the results of time-dependent Hartree–Fock calculations for the case of a 2D $H_2$. The binding potential is the same as given by Eq. (24), but with softening parameter 0.41 instead of 0.5. Additionally, the $H_2$ model assumes a mean-field potential

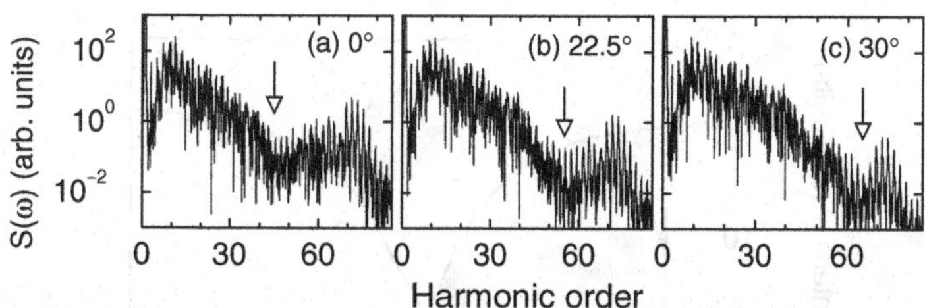


**Fig. 6.** Spectra of harmonics polarized parallel to the laser polarization direction for 2D $H_2$ in 780 nm laser pulses with intensity $5\times10^{14}$ W/cm$^2$. The alignment angle of the molecule is as indicated. From Ref. 15, copyright 2002 by the American Physical Society.

$$V_{ee}(\mathbf{r}) = \int \frac{|\Psi(\mathbf{r}',t)|^2 d^2r'}{\sqrt{(\mathbf{r}-\mathbf{r}')^2 + 0.36}} \tag{25}$$

describing the electron–electron interaction. The results for various orientations of this model molecule are shown in Fig. 6. Similar to $H_2^+$, the HHG spectra exhibit a minimum moving to higher harmonic orders with increasing angle between molecule and laser field.

HHG can be used for attosecond-pulse generation by superposition of harmonics with different frequencies. The duration and shape of attosecond pulses depend on the phases of the different harmonics. The harmonic phase can be calculated simply as the phase of the complex Fourier transformed acceleration, $\mathbf{a}(\omega)$. Figure 7 compares the orientation dependences of the harmonic intensity and of the harmonic phase for the 43rd harmonic in 2D $H_2^+$. It is apparent that the minimum in the harmonic yield coincides with a jump of the phase about $\pi$. Except for the jump, the phase is approximately constant. This suggests that the complex harmonic amplitude $\mathbf{a}(\omega)$ goes through zero at a certain angle of orientation.

If the observed minimum is due to destructive double-slit type interference, we can write the interference conditions in simple form. Assuming that the contributions from the two centers interfere with a phase difference determined by the de Broglie

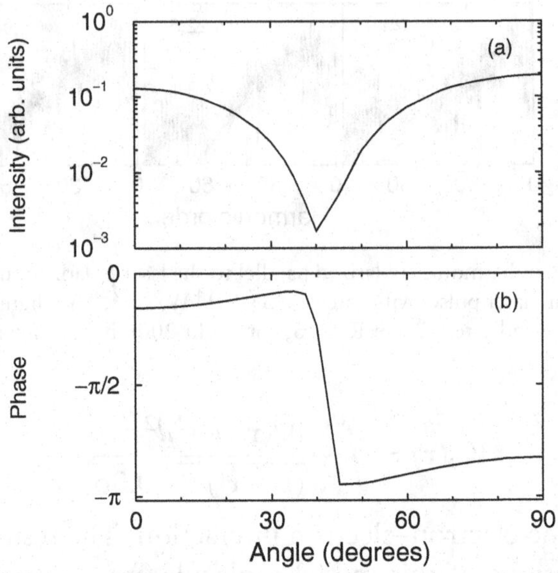


**Fig. 7.** Orientation dependence of the 43rd harmonic for 2D $H_2^+$ in a 780 nm laser pulse with intensity $5\times10^{14}$ W/cm$^2$. (a) Harmonic intensity; (b) harmonic phase. From Ref. 65, copyright 2002 by the American Physical Society.

wavelength $\lambda$ of the returning electron and the projection of the internuclear distance on the laser polarization axis, $R\cos\theta$, predicts interference minima at

$$R\cos\theta = (2n+1)\frac{\lambda}{2}, \quad n = 0, 1, 2, \ldots, \tag{26}$$

while interference maxima are expected for

$$R\cos\theta = n\lambda, \quad n = 1, 2, \ldots. \tag{27}$$

This model has recently been named the two-point emitter model. In order to corroborate the idea of two-center interference, data from extensive simulations of $H_2^+$ and $H_2$ was collected,[65] and the positions of the minima and maxima found in the spectra and in the orientation dependences were compared with the simple double-slit formulas. The result is plotted

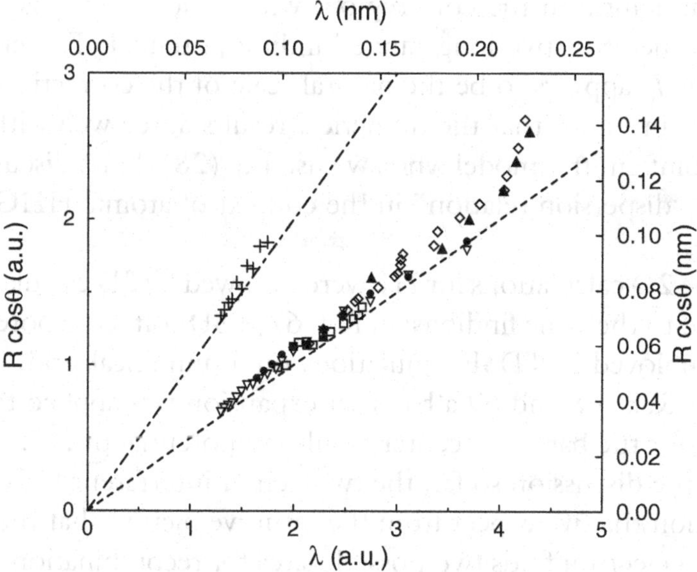


**Fig. 8.** Projection of the internuclear distance on the polarization axis, $R\cos\theta$, versus the de Broglie wavelength of the recolliding electron for which the harmonic yield is minimal (lower points) or maximal (upper points). The wavelength is calculated as explained in the text. Straight lines, two-point emitter model given by Eqs. (26) and (27). From Ref. 65, copyright 2002 by the American Physical Society.

in Fig. 8. The straight lines show the predictions of the formulas (26) and (27). The minima and maxima found from the numerical TDSE calculations are plotted as data points. For the conversion of harmonic frequencies $\omega$ to the electron wavelength $\lambda = 2\pi/k$, a heuristically corrected relation is used,

$$\frac{k^2}{2} = \omega, \tag{28}$$

instead of the usual relation $k^2/2 + I_\text{p} = \omega$ that emerges from the SFA or simple man's model. The idea behind this correction is that the interference is not determined by the wavelength corresponding to the asymptotic energy of the electron far away from the core, $E_\text{asympt} = \omega - I_\text{p}$. The interference is dictated by the wavelength in the core region, where the electron is faster due to the attractive long-range binding potential. The energy amount $I_p$ appears to be the natural scale of the correction. We observe in Fig. 8 that the numerical results agree well with the two-point emitter model when we use Eq. (28). For a discussion of this "dispersion relation" in the context of atomic HHG, see Ref. 66.

The 2D calculations for $H_2^+$ were followed by 3D calculations that led to the same findings: in Ref. 67, a 3D soft–core potential was employed in TDSE simulations on a numerical grid, while later in Refs. 68 and 69 a basis-set expansion was applied to the TDSE for the bare two-center Coulomb potential of $H_2^+$.

In the discussion so far, the two-center interference is a phenomenon that we expect from the intuitive picture that the recolliding electron finds two possible sites for recombination. One can derive the two-center interference from the SFA if one uses (i) the velocity form of the recombination matrix element and (ii) a linear combination of atomic orbitals (LCAO) for the initial electronic wave function, i.e.,

$$\Psi_0(\mathbf{r}) \sim \varphi_0(\mathbf{r} - \mathbf{R}/2) + \varphi_0(\mathbf{r} + \mathbf{R}/2). \tag{29}$$

One obtains for the matrix element

$$\langle \exp(i\mathbf{k}\cdot\mathbf{r})|i\nabla|\Psi_0\rangle \sim 2\cos(\mathbf{k}\cdot\mathbf{R}/2)\langle \exp(i\mathbf{k}\cdot\mathbf{r})|i\nabla|\varphi_0\rangle. \tag{30}$$

The function $\cos(\mathbf{k}\cdot\mathbf{R}/2)$ expresses explicitly the two-center interference and is consistent with the two-point emitter model, except for the different "dispersion relation" in the SFA. References 69 and 70 compare dipole and acceleration forms and show that the two-point emitter formulas follow from the acceleration form, if large internuclear distance is assumed as an additional approximation.

If the molecular orbital is an LCAO with opposite signs of the atomic orbitals, i.e.,

$$\Psi_0(\mathbf{r}) \sim \varphi_0(\mathbf{r} - \mathbf{R}/2) - \varphi_0(\mathbf{r} + \mathbf{R}/2), \tag{31}$$

one can show immediately that

$$\langle \exp(i\mathbf{k}\cdot\mathbf{r})|i\nabla|\Psi_0\rangle \sim 2i \sin(\mathbf{k}\cdot\mathbf{R}/2)\,\langle \exp(i\,\mathbf{k}\cdot\mathbf{r})|i\nabla|\varphi_0\rangle. \tag{32}$$

This means that the interference minima and maxima are now interchanged as compared to the case of Eq. (29).[65] It is important to note that the form of Eq. (31) does *not* imply that the orbital is antibonding or antisymmetric with respect to a mirror or inversion operation. Rather, the two atomic orbitals that are added with opposite signs in Eq. (31) are spatially translated relative to each other. This point has sometimes led to confusion in the literature; for clarification see also Refs. 71 and 72. An instructive example is the $\sigma_g$ valence orbital of the $N_2$ molecule. It is a symmetric orbital (with respect to both inversion and reflection), but in a very rough LCAO approximation, it would be the sum of two atomic p-orbitals, added with *opposite signs.* Therefore, the expected interference behavior is different from $H_2$ and $H_2^+$. The simple LCAO orbital, however, is not a realistic approximation for the $N_2$ orbital, since atomic s-orbitals contribute about 30% to the molecular orbital. The two s-orbitals have to be added with equal signs in order to give a symmetric molecular orbital. The two types of interference (corresponding to LCAOs with either plus or minus sign) are thus mixed together in one molecular orbital. As a result, there is no simple double-slit type interference in $N_2$.[73] This is confirmed by experiments on aligned $N_2$.[16,74]

Next, we consider the carbon dioxide molecule $CO_2$, which has the linear structure OCO. The doubly degenerate valence orbital has $\pi_g$ symmetry with two nodal planes, one along the molecular axis, and other perpendicular to it. The orbital can be well approximated by two atomic p-orbitals with opposite signs centered at the oxygen atoms. Therefore, two-center interference is expected for HHG from $CO_2$, and this is confirmed by experiments on aligned $CO_2$.[17,75] Kanai *et al.* have proved that the minimum observed in the HHG spectrum of $CO_2$ must be due to the recombination step in the three-step model by measuring simultaneously the harmonic intensity and the ionization yield. They found that the suppression of the harmonic intensity coincides with an increased ionization yield, ruling out the possibility that the suppression is caused by the ionization step.

Itatani *et al.*[16] proposed that the orbital dependence of HHG can be exploited to retrieve the orbital from the measured harmonics. This requires measurements of the HHG spectra for many different orientation angles of the molecule. One assumes that the HHG spectrum can be expressed as a product of a prefactor and the modulus squared of the recombination matrix element $\mathbf{d}(\omega, \theta)$,

$$S(\omega,\theta) \sim \omega^4 |a(\omega)\mathbf{d}(\omega,\theta)|^2. \tag{33}$$

By measuring the harmonic spectrum for a reference atom with known orbital, one can obtain the function $|a(\omega)|^2$. Assuming that the ionization and acceleration steps in the molecule are similar to the atom, one uses the same prefactor for the molecule so that from the measured molecular HHG spectra and Eq. (33), one obtains $|\mathbf{d}(\omega,\theta)|^2$. Assuming or measuring the phase of $\mathbf{d}$, one knows the complex values $\mathbf{d}(\omega,\theta)$. One can then obtain the orbital by inverse Fourier transformation. In Ref. 16 the phase was assumed to behave according to the findings of Refs. 15 and 65, namely that a phase jump by $\pi$ occurs where the harmonic intensity exhibits a minimum (see Fig. 7). A direct measurement of the harmonic phase seems possible,[7,76,77] but one must then also know the phase of the quantity $a(\omega)$ to obtain the phase of the transition matrix element $\mathbf{d}$. Reference 16 presents the result of the tomographic procedure for the $N_2$ molecule and finds good qualitative agreement with an *ab initio* orbital. Further work discussing multielectron effects is found in Refs. 78 and 79. Application to other molecules is work in progress. Complications arise for orbitals with nodal planes for the reasons explained in the previous section. At present it seems that for such orbitals, one has to prescribe the nodal planes by hand in order to make the tomographic reconstruction work.

## 4. Dynamical Effects

In the preceding sections, we have ignored the nuclear motion in laser-driven molecules. This is expected to be a good approximation for many molecules (including, for example, $N_2$) in ultrashort few-cycle pulses because the vibrational and rotational period is long compared to the pulse duration. The only expected effect is that one should average the results over the distribution of internuclear distances and orientations. In a typical experiment, this is the distribution corresponding to the vibrational ground state and random alignment of the molecular axes. However, one can also employ a prepulse that creates a rotational or a vibrational wavepacket which would then determine the distribution at the time of arrival of the strong driving pulse generating the harmonics. Note that the coherent superposition of the complex harmonic amplitudes for different molecular orientations/geometries has to be taken, rather than an incoherent summation of harmonic intensities.[80,81]

The assumption of frozen nuclei does not hold for very light molecules. In particular, we expect that nuclear motion takes places during the action of the driving pulse for molecules with bound hydrogen atoms. This includes of course hydrogen molecules, but also many other important species such as water or methane. The rotational motion can still be considered frozen during the laser pulse, but the vibrational period can be comparable to the pulse duration. We focus here on the possibility that the ionization of the molecule creates a vibrational wavepacket in the molecular ion. If there is significant wavepacket motion between the ionization and recombination steps in the HHG process, we expect an influence on the HHG spectra. Since the vibrational wavepacket is launched together with the continuum electron wavepacket, we have a *correlated wavepacket motion.*

Effects of nuclear motion have been seen in several simulations of HHG, in which the TDSE for $H_2^+$ was solved numerically.[82–85] The first demonstration of the effect of correlated nuclear and electronic wavepackets in HHG, however, was seen in TDSE calculations for two-dimensional $H_2$ and $D_2$ model molecules,[86] incorporating one-dimensional vibrational motion and two-dimensional electron motion. Figure 9 shows the ratio of the harmonics in $D_2$ and $H_2$ as well as the ratio $T_2/H_2$ for two different laser wavelengths, namely 780 and 1200 nm. The molecules are aligned perpendicular to the electric field of the laser pulse, except for the stars in Fig. 9(a), which refer to randomly aligned molecules. We see clearly that the heavier isotopes generate harmonics more efficiently. For example, the ratio

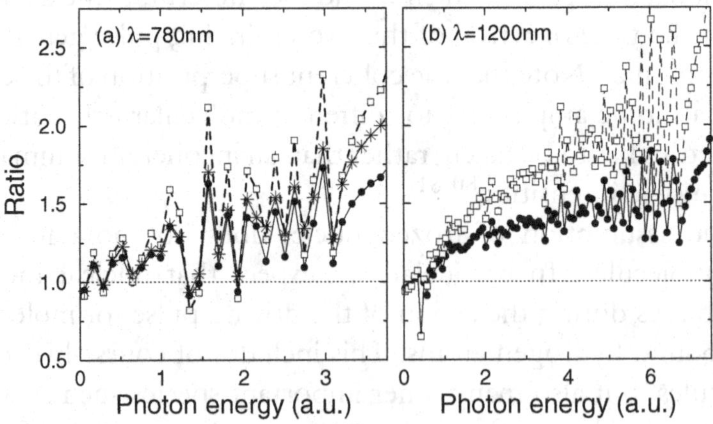


**Fig. 9.** Ratio of harmonics in different isotopes for 780 and 1200 nm laser pulses as indicated. Full circles, ratio $D_2/H_2$; open squares, ratio $T_2/H_2$ (molecules aligned perpendicular to the field). Stars in the left panel, ratio $D_2/H_2$ calculated for random alignment. From Ref. 86, copyright 2005 by the American Physical Society.

$D_2/H_2$ is almost always greater than one. The ratio tends to increase as a function of harmonic order. At least for this set of parameters, there is not a big difference between perpendicular and random alignment. We can understand the result in terms of the three-step model. In the time between ionization and recombination, the ionized molecule expands. This occurs faster in the lighter isotope $H_2$ because of the smaller reduced nuclear mass. Recombination must lead back into the initial state, i.e., into the vibronic ground state with internuclear distances near the equilibrium distance. The probability of this transition is reduced when the mean internuclear distance of the wavepacket in the molecular ion becomes larger. The suppression is therefore more pronounced in the lighter isotope. We can also interpret the increase of the ratio with harmonic order, since according to the three-step model, the higher harmonics are generated by longer electron travel times, see Eq. (9), leaving more time for the nuclear wavepacket dynamics. The physical process is illustrated in Fig. 10. It shows the ground-state Born–Oppenheimer (BO) potentials of $H_2$ and $H_2^+$ and the vibrational wavepacket dynamics in the $H_2^+$ potential.

To understand this effect quantitatively, we introduce the strong-field approximation for molecules including the vibrational dynamics of the molecular ion after ionization.[86–88] One obtains the time-dependent dipole moment

$$\begin{aligned}\mathbf{D}(t) = -2i \int_0^t dt' \left[\frac{2\pi}{i(t-t')+\epsilon}\right]^{3/2} \int_0^\infty dR\,\chi_0^*(R)\\ \times\, \mathbf{d}_\mathrm{r}^*(\mathbf{p}_\mathrm{s}(t',t)+\mathbf{A}(t))\exp\left(-\,iS(t',t)\right)\hat{U}_R(t-t')\\ \times\, d_i(\mathbf{p}_\mathrm{s}(t',t)+\mathbf{A}(t'),t')\chi_0(R) + \text{c.c.},\end{aligned} \tag{34}$$

where the saddle-point approximation for the integration over momenta has already been carried out. This equation corrects typographical errors in Eq. (3) of Ref. 88. $\chi_0(R)$ denotes the vibrational ground state of $H_2$, and the ionization and

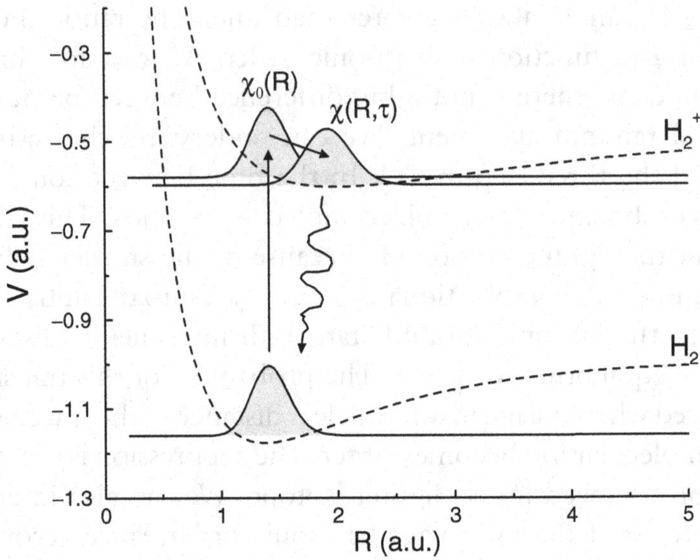


**Fig. 10.** Illustration of the harmonic-generation process in $H_2$. Shown are the Born–Oppenheimer potentials for the electronic ground states of $H_2$ and $H_2^+$ and the wavepacket evolution in the $H_2^+$ potential after ionization of $H_2$.

recombination matrix elements are

$$d_i(\mathbf{p}, t) = \langle \psi_{\mathrm{PW}}(\mathbf{p})\psi_R^+ | E(t) z | \psi_R \rangle_{\mathbf{rr}'}, \tag{35}$$

$$\mathbf{d}_{\mathrm{r}}(\mathbf{p}) = \langle \psi_{\mathrm{PW}}(\mathbf{p})\psi_R^+ | -\mathbf{r} | \psi_R \rangle_{\mathbf{rr}'}, \tag{36}$$

with $\psi_R(\mathbf{r}, \mathbf{r}')$ and $\psi_R^+(\mathbf{r}')$ being the electronic BO states in $H_2$ and $H_2^+$. The time-evolution operator $\hat{U}_R(t - t')$ propagates the vibrational wavepacket in the BO potential of the molecular ion, according to the one-dimensional TDSE for the internuclear distance. The prefactor 2 in Eq. (34) stands for the two electrons in $H_2$. Exchange terms[78,79,89] due to the two-electron nature of the system are neglected in Eq. (34).

If one could neglect the dependence of the matrix elements on $R$, the only dependences on $R$ in the SFA integral would be in $\chi_0(R)$ and in the time-evolution operator. This leads a simple expression for the SFA integral where only the vibrational autocorrelation function

$$C(\tau) = \int_0^\infty dR\, \chi^*(R,0)\chi(R,\tau) \qquad (37)$$

appears in addition to the SFA expression for atoms.[86] The autocorrelation function is the overlap of the evolved vibrational wavepacket at the time of recombination with the initial wavepacket at the time of ionization. If a harmonic is considered to be generated by only one pair of ionization time $t'$ and recombination time $t$, then the harmonic intensity is proportional to the modulus squared of the autocorrelation function, shown in Fig. 11 as a function of the travel time. The maxima at 18 fs for $H_2^+$ and 25 fs for $D_2^+$ correspond to the vibrational periods of these molecular ions. The typical electron travel times in HHG with 800 nm light, however, are below 2 fs since the optical period is only 2.7 fs. The inset shows $|C(\tau)|^2$ at such short times. In this region, the autocorrelation of $D_2^+$ is always greater than

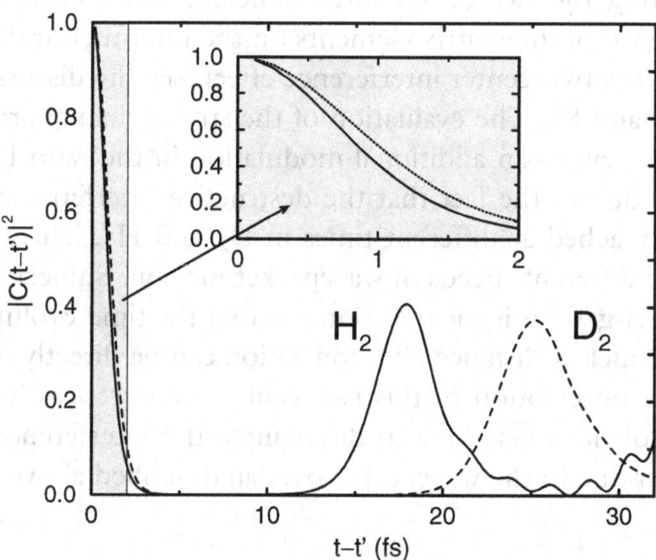


**Fig. 11.** Modulus squared of the vibrational autocorrelation functions for the wavepacket evolution in $H_2^+$ (solid) and $D_2^+$ (dashed). The initial wavepacket is the vibrational ground state of the neutral molecule.

the one of $H_2^+$, because of the faster dynamics in the latter. The ratio of the two curves is in good but not perfect agreement with the ratio of harmonics in $D_2$ and $H_2$ as calculated from the TDSE, see Fig. 3 in Ref. 86. The remaining discrepancies can be due to differences of the initial wavepacket from the vibrational ground state $\chi_0$, caused by the $R$-dependent ionization probability.[90] Another point is the slightly different ionization probabilities in $H_2$ and $D_2$, which can modify the harmonic ratios.[88]

The experiment of Ref. 18 confirms that $D_2$ produces more intense harmonics than $H_2$. Also, the experimental ratio increases as a function of harmonic order. In Refs. 18 and 86 it has been demonstrated that the time evolution of the internuclear distance can be reconstructed from the measured ratios by employing an iterative genetic algorithm that optimizes the time evolution such that the ratio of autocorrelation functions matches the measured ratio of harmonics.

The use of the simple autocorrelation function amounts to ignoring the two-center interference. The inclusion of the dependence of the matrix elements on the internuclear distance recovers the two-center interference effect, see the discussion in Refs. 18 and 88. The evaluation of the strong-field approximation then reveals an additional modulation in the ratio $D_2/H_2$ which is due to the fact that the destructive interference minimum is reached at different times in $D_2$ and $H_2$. This follows from the different speeds of wavepacket motion. Subject of current investigations is the question whether the time-evolution of the internuclear distance after ionization can be directly imaged from the observation of this two-center interference. It is not directly obvious yet, how to disentangle the interference from the effect due to the wavepacket overlap described above.

## 5. Conclusions

We have given an introduction into the physical phenomena taking place in high-order harmonic generation with molecules.

We have focused on three points: (i) the influence of molecular orbital structure on the ionization characteristics, (ii) the influence of the orbital on recombination (with two-center interference being the simplified perspective on this dependence in diatomic molecules), and (iii) the effects of vibrational motion on HHG. Molecular HHG is currently intensively studied by several experimental and theoretical groups, since it promises new approaches to molecular imaging, see also the review article of Ref. 90. Furthermore, molecules offer the possibility to manipulate the harmonic phase in a more flexible way than in atomic HHG. This may open new possibilities for attosecond-pulse shaping and for quasi phase matching.

## Acknowledgments

We thank J. Marangos and D. Villeneuve for valuable discussions, and we gratefully acknowledge discussions within the NSERC SRO network "Controlled electron re-scattering: femtosecond, sub-angstrom imaging of single molecules." This work was supported in part by the Deutsche Forschungsgemeinschaft.

## References

1. A. McPherson *et al.*, *J. Opt. Soc. Am. B* **4**, 595 (1987).
2. A. L'Huillier, K. J. Schafer and K. C. Kulander, *J. Phys. B* **24**, 3315 (1991).
3. T. Baeva, S. Gordienko and A. Pukhov, *Phys. Rev. E* **74**, 046404 (2006).
4. K. C. Kulander, K. J. Schafer and J. L. Krause, in *Proceedings of the Workshop, Super Intense Laser Atom Physics (SILAP) III*, edited by B. Piraux, A. L'Huillier and J. Rzążewski (Plenum Press, New York) (1993).
5. P. B. Corkum, *Phys. Rev. Lett.*, **71**, 1994 (1993).
6. P. M. Paul, E. S. Toma, P. Breger, G. Mullot, F. Augé, P. Balcou, H. G. Muller and P. Agostini, *Science* **292**, 1689 (2001).
7. Y. Mairesse *et al.*, *Science* **302**, 1540 (2003).
8. P. Tzallas, D. Charalambidis, N. A. Papadogiannis, K. Witte and G. D. Tsakiris, *Nature* **426**, 267 (2003).
9. M. Drescher, M. Hentschel, R. Kienberger, G. Tempea, C. Spielmann, G. A. Reider, P. B. Corkum and F. Krausz, *Science* **291**, 1923 (2001).

10. M. Hentschel *et al.*, *Nature* **414**, 509 (2001).
11. P. Agostini and L. F. DiMauro, *Rep. Prog. Phys.* **67**, 813 (2004).
12. A. Scrinzi, M. Y. Ivanov, R. Kienberger and D. M. Villeneuve, *J. Phys. B* **39**, R1 (2006).
13. G. Sansone *et al.*, *Science* **314**, 443 (2006).
14. A. H. Zewail, *J. Phys. Chem. A* **104**, 5660 (2000).
15. M. Lein, N. Hay, R. Velotta, J. P. Marangos and P. L. Knight, *Phys. Rev. Lett.* **88**, 183903 (2002).
16. J. Itatani *et al.*, *Nature* **432**, 867 (2004).
17. T. Kanai, S. Minemoto and H. Sakai, *Nature* **435**, 470 (2005).
18. S. Baker *et al.*, *Science* **312**, 424 (2006).
19. R. Velotta, N. Hay, M. B. Mason, M. Castillejo and J. P. Marangos, *Phys. Rev. Lett.* **87**, 183901 (2001).
20. J. Itatani, D. Zeidler, J. Levesque, M. Spanner, D. M. Villeneuve and P. B. Corkum, *Phys. Rev. Lett.* **94**, 123902 (2005).
21. H. Niikura *et al.*, *Nature* **417**, 917 (2002).
22. H. Niikura *et al.*, *Nature* **421**, 826 (2003).
23. F. Bloch and A. Nordsieck, *Phys. Rev.* **52**, 54 (1937).
24. W. Pauli and M. Fierz, *Nuovo Cimento* **15**, 167 (1938).
25. H. A. Kramers, *Collected Scientific Papers* (North Holland, Amsterdam, 1956).
26. W. C. Henneberger, *Phys. Rev. Lett.* **21**, 838 (1968).
27. F. H. M. Faisal, *J. Phys. B* **6**, L89 (1973).
28. B. Sundaram and P. W. Milonni, *Phys. Rev. A* **41**, 6571 (1990).
29. J. H. Eberly and M. V. Fedorov, *Phys. Rev. A* **45**, 4706 (1992).
30. K. Burnett, V. C. Reed, J. Cooper and P. L. Knight, *Phys. Rev. A* **45**, 3347 (1992).
31. D. G. Lappas, M. V. Fedorov and J. H. Eberly, *Phys. Rev. A* **47**, 1327 (1993).
32. M. Lewenstein, P. Balcou, M. Y. Ivanov, A. L'Huillier and P. B. Corkum, *Phys. Rev. A* **49**, 2117 (1994).
33. P. Salières, A. L'Huillier and M. Lewenstein, *Phys. Rev. Lett.* **74**, 3776 (1995).
34. P. Antoine, A. L'Huillier and M. Lewenstein, *Phys. Rev. Lett.* **77**, 1234 (1996).
35. P. Balcou, P. Salières, A. L'Huillier and M. Lewenstein, *Phys. Rev. A* **55**, 3204 (1997).
36. D. M. Wolkow, *Z. Physik* **94**, 250 (1935).
37. C. C. Chirilă and M. Lein, *Phys. Rev. A* **73**, 023410 (2006).
38. C. C. Chirilă and M. Lein, *J. Mod. Opt.* **54**, 1039 (2007).
39. D. Bauer, D. B. Milošević and W. Becker, *Phys. Rev. A* **72**, 023415 (2005).
40. T. K. Kjeldsen and L. B. Madsen, *J. Phys. B* **37**, 2033 (2004).
41. O. E. Alon, V. Averbukh and N. Moiseyev, *Phys. Rev. Lett.* **80**, 3743 (1998).
42. T. Kreibich, M. Lein, V. Engel and E. K. U. Gross, *Phys. Rev. Lett.* **87**, 103901 (2001).
43. A. Talebpour, C.-Y. Chien and S. L. Chin, *J. Phys. B* **29**, L677 (1996).
44. A. Talebpour, S. Larochelle and S. L. Chin, *J. Phys. B* **31**, L49 (1998).

45. C. Guo, M. Li, J. P. Nibarger and G. N. Gibson, *Phys. Rev. A* **58**, R4271 (1998).
46. E. Wells, M. J. DeWitt and R. R. Jones, *Phys. Rev. A* **66**, 013409 (2002).
47. S. M. Hankin, D. M. Villeneuve, P. B. Corkum and D. M. Rayner, *Phys. Rev. Lett.* **84**, 5082 (2000).
48. E. P. Benis *et al.*, *Phys. Rev. A* **70**, 025401 (2004).
49. M. J. DeWitt and R. J. Levis, *J. Chem. Phys.* **108**, 7739 (1998).
50. A. Saenz, *J. Phys. B* **33**, 4365 (2000).
51. C. Guo, *Phys. Rev. Lett.* **85**, 2276 (2000).
52. J. Muth-Böhm, A. Becker and F. H. M. Faisal, *Phys. Rev. Lett.*, **85**, 2280 (2000).
53. E. E. B. Campbell, K. Hoffmann, H. Rottke and I. V. Hertel, *J. Chem. Phys.* **114**, 1716 (2001).
54. M. Tchaplyguine *et al.*, *J. Chem. Phys.* **112**, 2781 (2000).
55. V. R. Bhardwaj, P. B. Corkum and D. M. Rayner, *Phys. Rev. Lett.* **91**, 203004 (2003).
56. D. Zeidler, A. B. Bardon, A. Staudte, D. M. Villeneuve, R. Dörner and P. B. Corkum, *J. Phys. B* **39**, L159 (2006).
57. M. Lein, *J. Phys. B* **36**, L155 (2003).
58. M. Lein, E. K. U. Gross and V. Engel, *J. Phys. B* **33**, 433 (2000).
59. M. D. Feit, J. A. Fleck Jr. and A. Steiger, *J. Comput. Phys.* 47, 412 (1982).
60. G. Lagmago Kamta and A. D. Bandrauk, *Phys. Rev. A* **74**, 033415 (2006).
61. R. Fischer, M. Lein and C. H. Keitel, *Phys. Rev. Lett.* **97**, 143901 (2006).
62. R. Fischer, M. Lein and C. H. Keitel, *J. Phys. B* **40**, F113 (2007).
63. M. W. Walser, C. H. Keitel, A. Scrinzi and T. Brabec, *Phys. Rev. Lett.* **85**, 5082 (2000).
64. N. J. Kylstra, R. M. Potvliege and C. J. Joachain, *J. Phys. B* **34**, L55 (2001).
65. M. Lein, N. Hay, R. Velotta, J. P. Marangos and P. L. Knight, *Phys. Rev. A* **66**, 023805 (2002).
66. J. Levesque, D. Zeidler, J. P. Marangos, P. B. Corkum and D. M. Villeneuve, *Phys. Rev. Lett.* **98**, 183903 (2007).
67. M. Lein, P. P. Corso, J. P. Marangos and P. L. Knight, *Phys. Rev. A* **67**, 023819 (2003).
68. G. Lagmago Kamta and A. D. Bandrauk, *Phys. Rev. A* **70**, 011404(R) (2004).
69. G. Lagmago Kamta and A. D. Bandrauk, *Phys. Rev. A* **71**, 053407 (2005).
70. A. Gordon and F. X. Kärtner, *Phys. Rev. Lett.* **95**, 223901 (2005).
71. T. K. Kjeldsen and L. B. Madsen, *Phys. Rev. A* **73**, 047401 (Comment) (2006).
72. V. I. Usachenko, *Phys. Rev. A* **73**, 047402 (Reply) (2006).
73. B. Zimmermann, M. Lein and J. M. Rost, *Phys. Rev. A* **71**, 033401 (2005).
74. P. Salières *et al.*, private communication.
75. C. Vozzi *et al.*, *Phys. Rev. Lett.* **95**, 153902 (2005).
76. H. Wabnitz *et al.*, *Eur. Phys. J. D* **40**, 305 (2006).
77. T. Kanai, E. J. Takahashi, Y. Nabekawa and K. Midorikawa, *Phys. Rev. Lett.* **98**, 153904 (2007).

78. S. Patchkovskii, Z. Zhao, T. Brabec and D. M. Villeneuve, *Phys. Rev. Lett.* **97**, 123003 (2006).
79. S. Patchkovskii, Z. Zhao, T. Brabec and D. M. Villeneuve, *J. Chem. Phys.* **126**, 114306 (2007).
80. M. Lein, R. de Nalda, E. Heesel, N. Hay, E. Springate, R. Velotta, M. Castillejo, P. L. Knight and J. P. Marangos, *J. Mod. Opt.* **52**, 465 (2005).
81. C. B. Madsen, A. S. Mouritzen, T. K. Kjeldesn and L. B. Madsen, *Phys. Rev. A* **76**, 035401 (2007).
82. S. Chelkowski, T. Zuo, O. Atabek and A. D. Bandrauk, *Phys. Rev. A* **52**, 2977 (1995).
83. K. C. Kulander, F. H. Mies and K. J. Schafer, *Phys. Rev. A* **53**, 2562 (1996).
84. W. Qu, Z. Chen, Z. Xu and C. H. Keitel, *Phys. Rev. A* **65**, 013402 (2001).
85. B. Feuerstein and U. Thumm, *Phys. Rev. A* **67**, 063408 (2003).
86. M. Lein, *Phys. Rev. Lett.* **94**, 053004 (2005).
87. C. C. Chirilă and M. Lein, *J. Mod. Opt.* **53**, 113 (2006).
88. C. C. Chirilă and M. Lein, *J. Phys. B* **39**, S437 (2006).
89. R. Santra and A. Gordon, *Phys. Rev. Lett.* **96**, 073906 (2006).
90. X. Urbain *et al.*, *Phys. Rev. Lett.* **92**, 163004 (2004).
91. M. Lein, *J. Phys. B* **40**, R135 (2007).

# Chapter 3

# Molecular Manipulation Techniques and Their Applications

Hirofumi Sakai

*Department of Physics, Graduate School of Science, The University of Tokyo, 7-3-1, Hongo, Bunkyo-ku, Tokyo 113-0033, Japan*
*hsakai@phys.s.u-tokyo.ac.jp*

A sample of aligned or oriented molecules is an ideal anisotropic quantum system. In this chapter, focuses are placed on the present status of molecular orientation based on laser technologies and applications with a sample of aligned molecules. After the introduction, the principle of molecular orientation with combined electrostatic and intense, nonresonant laser fields is outlined. In Secs. 3.1 and 3.2, it is shown how to achieve and observe one- and three-dimensional molecular orientation, respectively, with combined electrostatic and pulsed, nonresonant laser fields. In order to demonstrate the efficacy of aligned molecules, two representative works from our laboratory are presented in Sec. 4. In fact, a sample of aligned molecules is used to investigate multiphoton ionization processes in $I_2$ molecules with time-dependent polarization pulses (Sec. 4.1) and high-order harmonic generation from $N_2$, $O_2$, and $CO_2$ molecules (Sec. 4.2). Finally, some future subjects are presented in Sec. 5.

## 1. Introduction

The techniques of molecular manipulation based on laser technologies provide us with an ideal anisotropic quantum system. In fact, a sample of aligned or oriented molecules allows us to investigate quantum phenomena associated with molecular symmetry and correlations between a molecular axis and a laser polarization as well as stereodynamics in chemical reactions.[1] Here, "alignment" designates a state where molecular axes are parallel to each other but no attention is paid to the "head-versus-tail" order. On the other hand, "orientation" refers to a state where not only the axes are parallel to each other but also the heads of the molecules are directed in the same way. Some important physical phenomena such as high-order harmonic generation (HHG), nonsequential double ionization, and above-threshold ionization are caused by electron recollison.[2] Molecular alignment and orientation techniques are crucially important to explore "electronic stereodynamics" in molecules as suggested by Herschbach.[3]

Among various theoretical proposals to achieve molecular alignment or orientation, one of the most promising approaches is based on the interaction between an intense nonresonant laser field and an induced dipole moment of a molecule. In fact, theoretical works by Friedrich and Herschbach[4,5] were followed by experimental demonstrations of one-dimensional alignment with a linearly polarized laser field[6,7] and three-dimensional alignment with an elliptically polarized laser field[8] in the adiabatic regime where alignment proceeds slowly compared to the rotational period of a molecule. In the nonadiabatic regime, molecular alignment is realized as (fractional) revivals after exciting rotational wavepackets with ultrashort laser pulses.[9,10] As for molecular alignment in both adiabatic and nonadiabatic regimes, readers are referred to Refs. 11 and 12 and references therein.

Concerning molecular orientation, Friedrich and Herschbach pointed out that orientation in an electrostatic field can be greatly enhanced by applying an intense nonresonant laser field.[13,14] Based on this approach, the Tokyo group led by the author has given demonstrations of both one-dimensional[15–17] and three-dimensional orientations,[18] which are described in Secs. 3.1 and 3.2, respectively, after outlining the theoretical background in Sec. 2. Three-dimensional orientation corresponds to the complete directional control of molecules in the sense that all of three

Euler angles of a molecule are fixed in the laboratory frame with the "head-versus-tail" directions arranged the same way.

With a sample of aligned molecules, there have already been several successful investigations including multiphoton ionization and HHG. In fact, the effects of laser polarization on multiphoton ionization in aligned $I_2$ molecules have been examined with time-dependent polarization pulses,[19] which is described in Sec. 4.1. Quantum interference of electron de Broglie waves in the recombination process of HHG has been demonstrated with a sample of aligned $CO_2$ molecules,[20] which is described in Sec. 4.2. Other representative demonstrations using aligned molecules are the following two works. The branching ratio of photodissociation has been controlled using aligned $I_2$ molecules.[21] The three-dimensional structure of a single molecular orbital has been imaged by using HHG from aligned $N_2$ molecules.[22] Finally, some future subjects are presented in Sec. 5.

## 2. Theoretical Background

Here, the outline of the principle of molecular orientation with combined electrostatic and intense, nonresonant laser fields is described. More detailed discussions are found in Refs. 12–14. When a molecule is exposed to both an electrostatic field and a laser field, the Hamiltonian $H$ is expressed as the sum of rotational kinetic energy $H_{\text{rot}}$ and the interaction potential between the molecule and the fields $H_{\text{int}}$; $H = H_{\text{rot}} + H_{\text{int}}$. The interaction potential $H_{\text{int}}$ is the sum of the permanent dipole interaction potential $V_\mu = -\boldsymbol{\mu}\cdot \boldsymbol{E}_{\text{s}}$ and the anisotropic polarizability interaction potential, i.e., the induced dipole interaction potential $V_\alpha = -\sum_{\rho,\rho'} E_\rho \alpha_{\rho,\rho'} E_{\rho'}/4$. Here, $\boldsymbol{E}_{\text{s}}$ represents the electrostatic field; $\boldsymbol{\mu}$, the permanent dipole moment; $\boldsymbol{E}_{\rho(\rho')}$, the laser field components along the space-fixed Cartesian coordinate axes ($\rho, \rho' = \text{x, y, z}$), and $\alpha$ is the polarizability tensor of the molecule.

The induced dipole interaction with a linearly polarized laser field produces a double-well potential along the polarization direction, where the molecular axis librates around the polarization direction. The pendular energy levels form nearly degenerate tunneling doublet states of opposite parity. If the molecule is polar, the introduction of an electrostatic field couples the components of a given tunneling doublet. Thus, even a relatively weak electrostatic field can convert second-order alignment by a laser field into a strong first-order orientation.

On the other hand, it has been theoretically and experimentally shown that molecules can be three-dimensionally aligned with an elliptically polarized laser field, i.e., a molecular plane can be confined to the polarization plane with the most polarizable molecular axis parallel to the major axis of the elliptical polarization.[8] Three-dimensional molecular orientation is based on the combination of one-dimensional orientation and three-dimensional alignment. With the combined electrostatic and elliptically polarized laser fields, three-dimensional double-well potential should be produced with the asymmetry along the electrostatic field direction.

## 3. Molecular Orientation with Combined Electrostatic and Intense, Nonresonant Laser Fields

### 3.1. *One-dimensional molecular orientation*

In this section, it is shown how to achieve and observe one-dimensional molecular orientation.[15–17] Figure 1 shows the schematic diagram of the experimental setup. A pulsed supersonic beam of OCS molecules is supplied by expanding OCS molecules diluted (5%) with argon (Ar) or helium (He) gas through a 0.25-mm-diam nozzle. The molecular beam is parallel to the time-of-flight (TOF) axis and crossed at 90° by the focused laser pulses. To achieve molecular orientation, we use an extraction field of

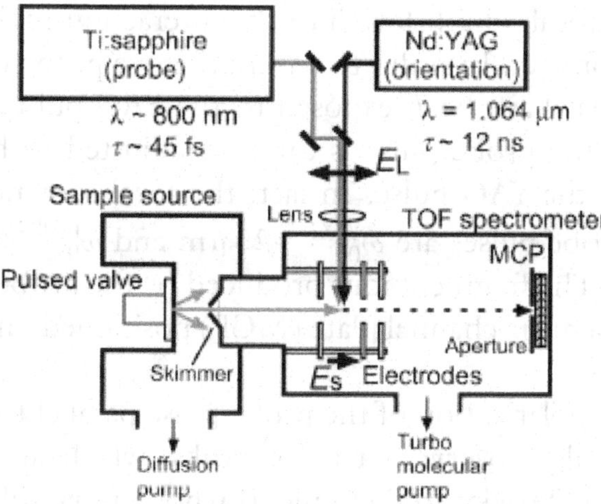


**Fig. 1.** A schematic diagram of the experimental setup. (Reprinted with permission from Ref. 16. Copyright (2003) by the American Institute of Physics.)

the TOF spectrometer for the interaction between the permanent dipole moment and the electrostatic field. The molecular orientation can be enhanced by the polarizability anisotropy interaction between the induced dipole moment and the intense, nonresonant laser pulse (wavelength $\lambda = 1064$ nm), which comes from an injection-seeded Nd:YAG laser. The polarization of the YAG pulse is parallel to that of the electrostatic field, i.e., the TOF axis. The maximum peak intensity used is $2.6 \times 10^{12}$ W/cm$^2$. The typical duration of the YAG pulse is ~12 ns (full width at half maximum) and long enough to ensure that the experiments are performed in an adiabatic regime where the orientation proceeds slowly compared to the rotational period of the molecule.

An intense femtosecond laser pulse ("probe" pulse) is used to ionize the OCS molecules at the peak of the YAG pulse to see the degree of orientation. The femtosecond laser pulses are delivered by a Ti:sapphire based amplified laser system and are centered at ~800 nm. The typical pulse duration is 45 fs and the peak intensity used is $3 \times 10^{14}$ W/cm$^2$. The laser pulses are spatially overlapped using a dichroic mirror and focused by a 30-cm-focal-length lens into the interaction region of the TOF spectrometer. In order to ensure that we probe only those molecules that have been exposed to the YAG pulse, the focal spot size of the probe pulse is carefully adjusted to be smaller than that of the YAG pulse. In fact, the focal spot radii of the YAG and probe pulses are $\omega_0^{\mathrm{YAG}} \sim 25\,\mu\mathrm{m}$ and $\omega_0^{\mathrm{probe}} \sim 14\,\mu\mathrm{m}$, respectively. The fragment ions produced by the probe pulse are detected by a microchannel plate (MCP) positioned on-axis with the TOF axis.

With the polarization of the probe pulse parallel to the TOF axis, we usually observe a pair of peaks, which are so-called "forward" and "backward" fragments whose initial velocities are directed toward and away from the MCP detector, respectively. This is the reflection of enhanced ionization by which molecules initially aligned along the polarization are much more easily to be multiply ionized. Figure 2 shows the typical TOF spectra observed (a) with and (b) without YAG pulses. Here, we concentrate our attention on the $S^{3+}$ ion signals. When the YAG pulses

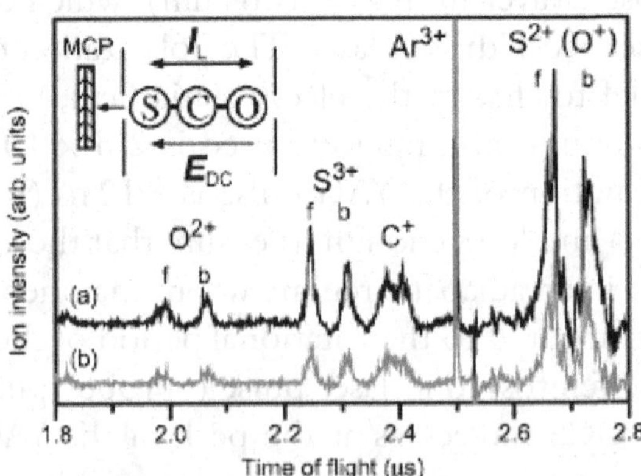


**Fig. 2.** Typical TOF spectra of OCS molecules (a) with and (b) without YAG pulses. The labels f and b denote fragment ions coming from the forward and backward initial emission directions, respectively. The inset shows the directions of the electrostatic field, the linearly polarized laser field, and an oriented OCS molecule. (Reprinted with permission from Ref. 15. Copyright (2003) by the American Physical Society.)

are not present (b), the forward and backward signals look almost symmetric, indicating that the molecules are randomly oriented. This observation also allows us to think that the probe pulse does not play any significant role for the orientation of OCS molecules under the present conditions.

When the YAG pulses are present (a), the signals look asymmetric. We interpret this observation as the result that more than half the OCS molecules are oriented with their S atoms directed toward the MCP detector. In fact, the change observed in Fig. 1(a) is also consistent with our expectation because the permanent dipole moment of an OCS molecule is directed toward the S atom and the molecules tend to be oriented in the way shown in the inset of Fig. 1. The enhancement of the backward signal is the reflection of the combined effect of molecular alignment and enhanced ionization, i.e., some molecules are not oriented but just aligned with the application of the YAG pulses. We can further confirm our interpretation by observing the signal magnitude of a counterpart fragment, e.g., $CO^+$. As expected, the background signal is larger than the forward signal when the YAG pulses are present.

The molecular orientation can be controlled by the peak intensity of the YAG pulse, the magnitude of the electrostatic field, and the rotational temperature of the molecules. The rotational temperature of the molecules can be controlled by the carrier gas and/or its backing pressure in the pulsed gas valve. For a fixed backing pressure, the Ar gas can supply a molecular sample with lower rotational temperature than the He gas. For a fixed carrier gas, lower rotational temperature can be achieved by higher backing pressures. Theoretically, the degree of orientation can be expected to increase by increasing the peak intensity of the YAG pulse and the magnitude of the electrostatic field or by lowering the rotational temperature of the molecules. These expectations have been experimentally confirmed.[15–17]

### 3.2. *Three-dimensional molecular orientation*

As demonstrated in Sec. 3.1, one-dimensional molecular orientation can be achieved with combined electrostatic and intense, nonresonant, linearly polarized laser fields. On the other hand, it has been theoretically and experimentally shown that molecules can be three-dimensionally aligned using an elliptically polarized laser field as described in Sec. 2.[8] As a corollary of these two facts, the combination of an electrostatic field and an elliptically polarized laser field should create three-dimensional potential wells with the asymmetry along the electrostatic field, which could be used to achieve three-dimensional molecular orientation.

Three-dimensional molecular orientation can be confirmed by complementary observations using two-dimensional ion imaging and TOF mass spectrometry.[18] The schematic diagrams of the two ways of observations are shown in Fig. 3. Three-dimensional alignment can be observed by two-dimensional ion images. Orientation is confirmed by observing TOF spectra as in one-dimensional molecular orientation described in Sec. 3.1.

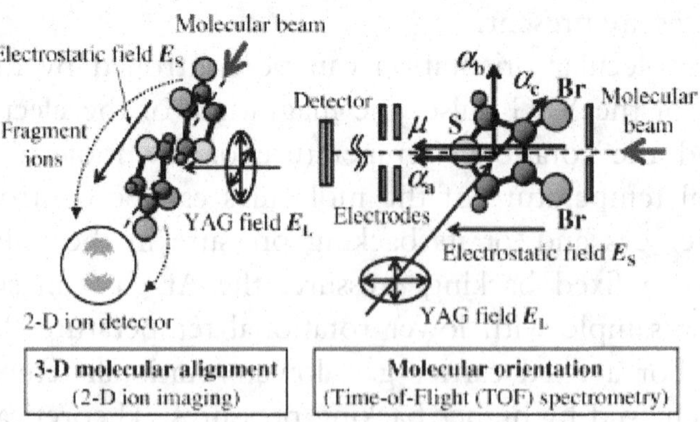


**Fig. 3.** Schematic diagrams of the complementary observations employed to confirm three-dimensional orientation. Three-dimensional alignment is confirmed with the two-dimensional ion imaging technique on the left and the one-dimensional molecular orientation is verified by TOF mass spectrometry on the right. $\mu$ and $\alpha$ represent the permanent dipole moment and the polarizability of 3,4-dibromothiophene molecule, respectively. (Reprinted with permission from Ref. 18. Copyright (2005) by the American Physical Society.)

These two ways of observations provide us with sure evidence of three-dimensional molecular orientation.

As a sample molecule, we employ 3,4-dibromothiophene (DBT) illustrated in Fig. 3. It has a nonzero permanent dipole moment and three different polarizability components, which are required of the demonstration of three-dimensional orientation. We concentrate our attention on $Br^+$ and $S^+$ ion signals for observing three-dimensional orientation.

In order to achieve three-dimensional orientation, the polarization of the YAG pulse is elliptical in the present experiments. Ellipticity is defined as $\varepsilon = E_{\rm min}/E_{\rm maj}$, where $E_{\rm maj}$ and $E_{\rm min}$ are the major and minor axis components of the laser electric field, respectively. The ellipticity is adjusted by rotating a half-wave plate placed before a fixed quarter-wave plate. Thereby, we can use arbitrary ellipticity with two axes of the ellipse fixed in the laboratory frame. The molecular beam is supplied by expanding DBT molecules seeded in He gas into a vacuum chamber through a 0.5-mm-diam nozzle of the pulsed gas valve. The partial pressure of DBT is 7 Torr and the backing pressure is changed from 1 to 8 atm. The applied voltages of the TOF spectrometer are adjusted for velocity mapping condition so that the fragment ions with the same initial velocity arrive at the same position of the detector plane. The rest of experimental details is the same as in the case of one-dimensional orientation.

First, we confirm the three-dimensional alignment achieved with an elliptically polarized laser field based on the observations of ion images. Figure 4 shows typical images of $Br^+$ and $S^+$ ions observed by using various ellipticities of the YAG pulse. Here, the probe pulse is circularly polarized in order to avoid any influence of enhanced ionization on ion images.

Figure 4(a) shows ion images observed without YAG pulses. They look almost isotropic, indicating that the molecules are randomly oriented. When YAG pulses ($3\times10^{12}$ W/cm$^2$) are present, distinct anisotropies are observed in the images (Figs. 4(b)–4(e)).

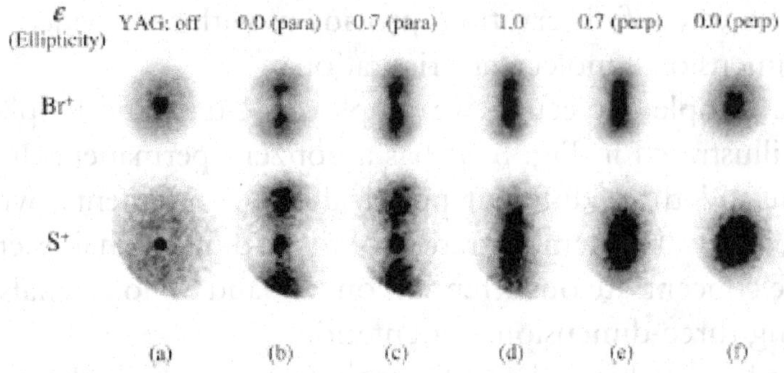


**Fig. 4.** Typical ion images of $Br^+$ (upper row) and $S^+$ (lower row) fragments observed with different ellipticities of the YAG pulses ($3 \times 10^{12}$ W/cm$^2$). The labels (para) and (perp) mean the major axis of the polarization parallel and perpendicular to the detector plane, respectively. (Reprinted with permission from Ref. 18. Copyright (2005) by the American Physical Society.)

When the YAG pulses are linearly polarized with the polarization parallel to the detector plane (Fig. 4(b)), both $Br^+$ and $S^+$ ions gather along the polarization and divided into top and bottom parts. Considering the structure of DBT illustrated in Fig. 3, we can interpret the change observed in Fig. 4(b) as the evidence that the major axis of the molecule defined as the axis with the largest polarizability component is confined along the polarization. Figure 4(f) observed with the polarization perpendicular to the detector plane can serve to complement Fig. 4(b). The isotropic distributions of both $Br^+$ and $S^+$ ions show that the major axis of the molecule is now perpendicular to the detector plane. Figure 4(d) observed with circular polarization shows that both ions are distributed along the vertical direction and the division into top and bottom parts has disappeared. Based on the theoretical expectations, we can interpret this observation as the result that the molecular plane comes to be confined in the plane of circular polarization but the molecule is allowed to rotate in the polarization plane. Figures 4(c) and 4(e) are observed with elliptical polarization ($\varepsilon = 0.7$) with the major axis of the ellipse parallel and perpendicular, respectively, to the detector plane. The images of $S^+$ ions shown in Figs. 4(c) and 4(e) indicate that the major axis of the molecule is confined parallel to that of elliptical polarization. On the other hand, the images of $Br^+$ ions are distributed along the vertical direction. When Fig. 4(e) is compared to Fig. 4(f), we notice that the effect of elliptical polarization is distinct in squeezing the image of $Br^+$ ions in the horizontal direction and extending in the vertical direction. A series of images shown in Fig. 4 gives us clear evidence that, under the presence of an elliptically polarized laser field, the molecules are confined in the polarization plane with their major axes parallel to the major axis of the ellipse.

Next, we confirm the "head-versus-tail" order of the molecules when an electrostatic field is also present. For this purpose, we make the major axis of elliptical polarization perpendicular to the detector plane, i.e., parallel to the TOF axis. As described in Sec. 3.1, we use an extraction field of the TOF spectrometer to achieve molecular orientation. Here, we examine the $S^+$ ion signals in the TOF (Fig. 5) obtained with (black) and without (gray) the YAG pulses ($3 \times 10^{12}$ W/cm$^2$). In the

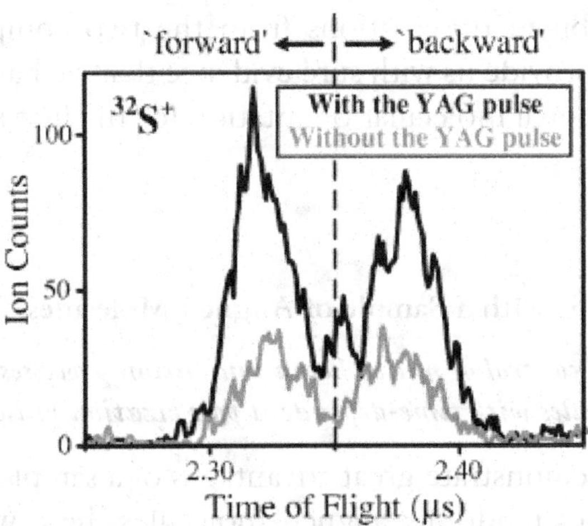


**Fig. 5.** The peaks of $S^+$ ions in the TOF spectra observed with (black) and without (gray) YAG pulses ($3 \times 10^{12}$ W/cm$^2$). The electrostatic field and the backing pressure are 760 V/cm and 8 atm, respectively. (Reprinted with permission from Ref. 18. Copyright (2005) by the American Physical Society.)

TOF spectrum obtained without the YAG pulses, the two peaks look almost symmetric, indicating that the molecules are randomly oriented. This also ensures that nonadiabatic orientation induced by the intense femtosecond probe pulse is negligible under the present conditions. Considering the permanent dipole moment of DBT illustrated in Fig. 3, the S atom of DBT should be directed toward the detector and the forward $S^+$ fragments should be increased if orientation is achieved. As expected, the TOF spectrum shows distinct asymmetry when the YAG pulses are present. The intensity of the forward signal is more enhanced than that of the backward one, indicating that more than half the molecules are oriented with the S atoms directed toward the detector. This observation can be reasonably interpreted as the evidence of molecular orientation.

As in the case of one-dimensional molecular orientation described in Sec. 3.1, we confirm that the degree of orientation can be enhanced when the peak intensity of the YAG pulse and the magnitude of the electrostatic field[23] are increased or the rotational temperature of the molecules is lowered.

The combined observations from the two complementary experiments provide us with sure evidence that we have achieved three-dimensional molecular orientation for the first time to our knowledge.

## 4. Applications with a Sample of Aligned Molecules

### 4.1. *Optimal control of multiphoton ionization processes in aligned $I_2$ molecules with time-dependent polarization pulses*

In order to demonstrate great advantages of a sample of aligned molecules over randomly aligned molecules, here we describe our pioneering experiment on optimal control of multiphoton ionization processes in aligned $I_2$ molecules with time-dependent polarization pulses. There are two remarkable features in this experiment. One is that a sample of aligned molecules is employed in the optimal control experiments for the first time. In fact, a sample of aligned molecules allows us to investigate correlations between laser polarization and the molecular axis. The other is that a time-dependent polarization pulse is used as a new control parameter. By using time-dependent polarization pulses, we can make the best use of vectorial properties of a laser electric field.[24]

Here, we focus on the multiphoton ionization processes in $I_2$ molecules. It is well known that nonsequential double ionization is caused by the recolliding electron.[2] Therefore, if the polarization deviates very much from linear, when the electron returns to the parent ion, it is offset in the lateral direction and it can miss. As a corollary of nonsequential double ionization, we can expect that it should contribute more to the production of doubly charged molecular ions $I_2^{2+}$ than that of oddly charged molecular ions such as ($I_2^+$ and) $I_2^{3+}$.

We investigate the correlation experimentally. Figure 6 shows the experimental setup. A pulsed supersonic beam of $I_2$ molecules is introduced into the TOF mass spectrometer by expanding $I_2$ molecules (~1 Torr) buffered with Ar gas (760 Torr) through a 0.25-mm-diam nozzle. The $I_2$ molecules are aligned by the irradiation of pulses from an injection-seeded Nd:YAG laser as described in Sec. 3.1. The polarization is set to be parallel to the TOF axis and thereby the $I_2$ molecules are aligned along the TOF axis. Under the present conditions, the alignment cosine defined as $\langle\langle\cos^2\theta\rangle\rangle$, where $\theta$ is the polar angle between the molecular axis and the polarization direction, i.e., the TOF axis in this case, is estimated to be ~0.7. Intense femtosecond Ti:sapphire laser pulses are used to ionize the $I_2$ molecules.

Here, we introduce a fitness function $F$ defined by $F = I(1,1)/\,[I(1,0) + I(1,2) + I(2,1)]$, where $I(m,n)$ is the integrated $I^{m+}$ signals of the dissociation channel $I^{m+} + I^{n+}$ produced from $I_2^{(m+n)+}$. In general, higher laser intensities are

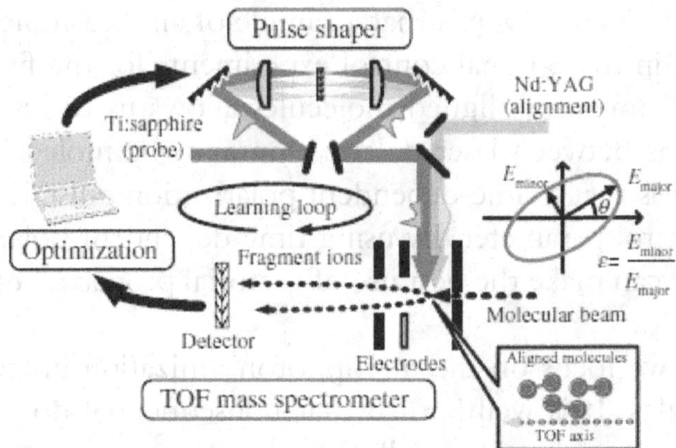


**Fig. 6.** A schematic diagram of the experimental setup. The pulse shaper is used to generate shaped femtosecond pulses with time-dependent polarizations. Experimental feedback signals are collected from TOF spectra that result from the interaction of the shaped femtosecond pulses with aligned molecules. These signals are processed in a computer to iteratively improve the time-dependent polarization pulses based on the genetic algorithm until an optimal result is obtained. (Reprinted with permission from Ref. 19. Copyright (2004) by the American Physical Society.)

necessary to produce more highly charged molecular ions. In order to reduce such an intensity-dependent effect on the fitness $F$, we compare ion signals from both $I_2^+$ and $I_2^{3+}$ (oddly charged molecular ions) with those from doubly charged molecular ions $I_2^{2+}$.

We measure the ellipticity dependence of the fitness $F$ by rotating a half-wave plate placed before a fixed quarter-wave plate. Ion signals are accumulated with 2000 laser shots for each ellipticity and the results are shown in Fig. 7. The fitness $F$ is 0.215 for linear polarization and gradually decreases down to 0.195 for circular polarization. This means that the relative production efficiency of doubly charged molecular ions $I_2^{2+}$ is high for linear polarization and that of oddly charged molecular ions ($I_2^+$ and $I_2^{3+}$) is high for circular polarization, which is qualitatively consistent with our expectation. The contrast defined by (the maximum fitness $F$)/(the minimum fitness $F$) is ~1.1. When molecules are randomly oriented, no significant

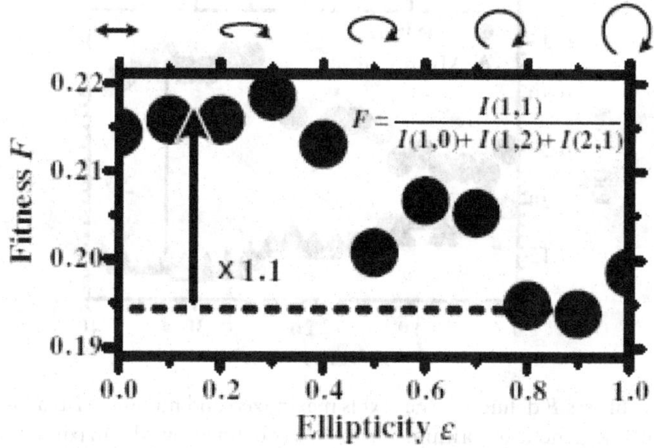


**Fig. 7.** Ellipticity dependence of the fitness $F$ defined in the text. The contrast defined by (the maximum $F$)/(the minimum $F$) is $\sim$ 1.1. (Reprinted with permission from Ref. 19. Copyright (2004) by the American Physical Society.)

correlation between the fitness $F$ and the ellipticity $\varepsilon$ has been observed. This already demonstrates the efficacy of a sample of aligned molecules.

Then we optimize the fitness $F$ with the homemade learning-loop optimal control system with which time-dependent polarization pulses can be generated and controlled. We employ a typical $4f$ configuration pulse shaper but the polarizers are removed to generate time-dependent polarization pulses. In order to search for optimal time-dependent polarization pulses, the genetic algorithm (GA) is used as an optimization algorithm. The ellipticity for each frequency component is used as the only parameter in GA and the phase for each frequency component is kept constant. As an initial pulse, we use a randomly polarized pulse for which the ellipticity for each frequency component is randomly defined. With a sample of aligned molecules, TOF signals become sharper compared to those with randomly oriented molecules. Sharper and isolated signals are very advantageous to be used for evaluating the fitness $F$.

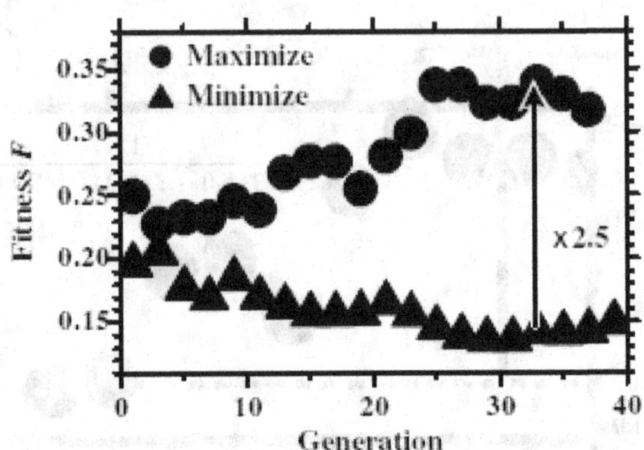


**Fig. 8.** The fitness $F$ defined in the text is maximized and minimized as a function of generation of the genetic algorithm. The contrast defined by (the maximum $F$)/(the minimum $F$) is $\sim 2.5$, which is much larger than that obtained with a fixed ellipticity as shown in Fig. 7. (Reprinted with permission from Ref. 19. Copyright (2004) by the American Physical Society.)

We maximize and minimize the fitness $F$ and its evolutions are shown in Fig. 8 as a function of generation of GA. The maximum (minimum) fitness obtained is 0.34 (0.14) and the contrast is ~2.5, which is much larger than that obtained with a fixed ellipticity (~1.1). Our results show a much better ability to control charged states of molecular ions with time-dependent polarization pulses than with elliptically polarized pulses having a fixed ellipticity. The optimal pulses obtained are characterized with the technique known as POLLIWOG (polarization labeled by interference versus wavelength of only a glint).[25] The great difference between the optimal pulses giving the maximum $F$ and those giving the minimum $F$ is revealed and discussed in our original paper.[19]

The present experiments point to the following new directions in optimal control studies with molecular systems: (a) A sample of aligned molecules is employed in the optimal control experiments for the first time. (b) In order to optimize multiphoton ionization processes in molecules, time-dependent polarization pulses are applied to the learning-loop optimal control system for the first time. (c) With a sample of aligned molecules and time-dependent polarization pulses, both external and internal degrees of freedom in molecules are simultaneously controlled.

Inspired by the present experiments, we have developed a new theoretical model to predict tunnel ionization probabilities of molecules with time-dependent polarization pulses.[26] The model is based on the combination of the molecular ADK theory[27] and Landau–Zener transitions. It has been shown that one can control tunnel ionization probabilities with appropriately designed time-dependent polarization pulses. In order to verify the validity of such theoretically predicted time-dependent polarization pulses, we have also developed a technique for the generation of those pulses based on the combination of a direct comparison between the target and the analyzed results of a polarization-characterization measurement and an adaptive learning loop with GA.[28]

### 4.2. *High-order harmonic generation from aligned molecules*

High-order harmonic generation (HHG) from atoms and molecules is one of central subjects in intense laser physics over decades because it has a potential for a (tunable) coherent ultrashort light source in the extreme ultraviolet and soft X-ray regions. HHG from atoms is well understood by the three-step model.[2] First, an electron tunnels through the atomic potential barrier modified by the intense laser field and appears in the continuum (step 1). Then the freed electron is driven by the intense laser field and has a probability of returning to the parent ion after the field reverses its direction (step 2). High-order harmonics can be emitted if the recollision with the parent ion leads to recombination (step 3). As long as a sample of randomly aligned molecules is employed, high-order harmonics from molecules have shown generation characteristics similar to those from atoms with the same ionization potential. However, a sample of aligned molecules gives us an opportunity to investigate quantum phenomena in HHG associated with molecular symmetries. In particular, simultaneous observations of both ion yields and harmonic signals under the same experimental conditions serve to disentangle the contributions from the ionization and recombination processes. In fact, clear evidence for quantum interference of electron de Broglie waves in the recombination process has been recently demonstrated using aligned $CO_2$ molecules.[20]

In the present experiment, an output from a Ti:sapphire based chirped pulse amplification system with a pulse width of ~50 fs and a center wavelength of ~800 nm is split into two pulses with a Michelson-type interferometer. The first pulse is used as a pump to create rotational wavepackets and induce nonadiabatic molecular alignment. The second pulse is delayed by a computer controlled translation stage and is used as a probe to generate high-order harmonics. Both of the two pulses are focused with a lens ($f = 300$ mm) into a supersonic gas jet in the vacuum chamber. The intensity of the pump pulse is $\sim 6 \times 10^{13}$ W/cm$^2$ and that of the probe pulse is $\sim 2 \times 10^{14}$ W/cm$^2$, which is well below the saturation intensity. The generated harmonics are spectrally resolved by a 1-m grazing incidence monochromator with a Pt-coated 300-grooves/mm grating and detected by an electron multiplier. The ions produced through multiphoton ionization are detected by a cylindrical ion collector located downstream of the gas jet.

Figure 9 shows the evolution of the ion yield (Fig. 9(a)) and the 23rd harmonics (Figs. 9(b) and 9(c)) from $N_2$ molecules as a function of the delay ($\tau$) between the pump and probe pulses. The polarizations of the pump and probe pulses are parallel (Figs. 9(a) and 9 (b)) or perpendicular (Fig. 9(c)) to each other. Both of the ion and harmonic intensities modulate at a period of ~2 ps, which corresponds to a quarter of

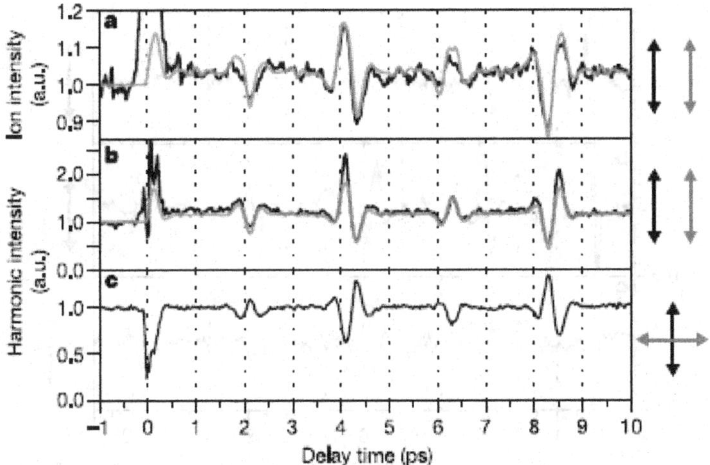


**Fig. 9.** The temporal evolution of the ion yield and the 23rd harmonic intensity from $N_2$ molecules. (a–c) The ion yield dominated by $N_2^+$ (a) and the 23rd harmonics from $N_2$ molecules (b, c) as a function of the pump–probe delay. The polarizations of the pump and probe pulses are parallel (a, b) or perpendicular (c) to each other. The rotational period $T_{rot}$ of a $N_2$ molecule is 8.4 ps. The results of theoretical calculations are also shown by gray curves. The rotational temperature of the $N_2$ molecules is assumed to be 80 K and the highest degree of alignment $\langle \cos^2 \theta \rangle$ is estimated to be 0.62 at $T_{rot}/2$. (Reprinted with permission from Ref. 20. Copyright (2005) by the Nature Publishing Group.)

the rotational period $T_{rot}$ ($\sim$8.4 ps) of neutral $N_2$ molecules. The modulations at every $T_{rot}/4$ are characteristic of those molecules whose highest occupied molecular orbitals (HOMOs) have $\sigma_g$ symmetry. As can be seen in Figs. 9(a) and 9(b), the harmonic intensity modulates in phase with the ion intensity when the polarizations are parallel to each other. Since the ion yields reflect the degree of alignment along the probe laser polarization, the correlation between the ionization and HHG indicates that $N_2$ molecules aligned along the probe laser polarization efficiently generate high-order harmonics, while the anti-aligned molecules suppress them as shown in Fig. 9(c). The results of theoretical calculations are also included in Fig. 9. We see satisfactory agreement between the experiments and the calculations.

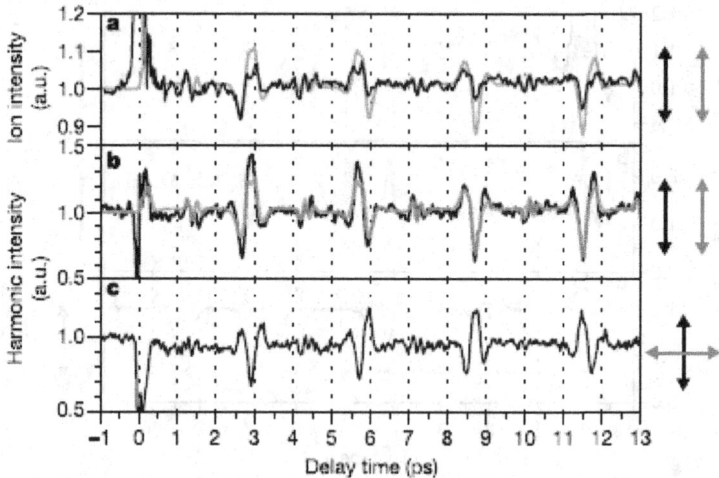


**Fig. 10.** The temporal evolution of the ion yield and the 23rd harmonic intensity from $O_2$ molecules. (a–c) The ion yield dominated by $O_2^+$ (a) and the 23rd harmonics from $O_2$ molecules (b, c) as a function of the pump–probe delay. The experimental conditions are identical with those of Fig. 9. The rotational period $T_{rot}$ of an $O_2$ molecule is 11.6 ps. The results of theoretical calculations are also shown by gray curves. The rotational temperature of the $O_2$ molecules is assumed to be 80 K and the highest degree of alignment $\langle \cos^2 \theta \rangle$ is estimated to be 0.62 at $T_{rot}/4$. (Reprinted with permission from Ref. 20. Copyright (2005) by the Nature Publishing Group.)

As a sample molecule whose HOMO has $\pi_g$ symmetry, we first investigate $O_2$ molecules. The experimental results are shown in Fig. 10 and are also in good agreement with the results of theoretical calculations. The modulations of both the ion and harmonic intensities for $O_2$ molecules (Fig. 10) are quite different from those for $N_2$ molecules (Fig. 9). Since $T_{rot} = 11.6$ ps for $O_2$, we see a modulation at every one-eighth of $T_{rot}$, which is characteristic of molecules whose HOMOs have anti-bonding $\pi_g$ symmetry. When the polarizations are parallel to each other, the harmonic intensity (Fig. 10(b)) modulates in phase with the ion intensity (Fig. 10(a)). On the other hand, when the polarizations are perpendicular to each other, the harmonic intensity (Fig. 10(c)) modulates out of phase with that in the parallel case (Fig. 10(b)). These correlations between the ionization and HHG are the same as those observed for $N_2$.

We further investigate the crucial role of the symmetry of HOMO in HHG using $CO_2$ molecules. The HOMO of a $CO_2$ molecule, which is dominated by the two $p$ orbitals in the two O atoms, has the same anti-bonding $\pi_g$ symmetry as that of an $O_2$ molecule. The difference between $O_2$ and $CO_2$ molecules is in the distance $R$ between the two O atoms; $R$ of $CO_2$ (0.232 nm) is about twice as long as $R$ of $O_2$ (0.121 nm). This difference in $R$ allows us to examine its influence on HHG. Figure 11 shows the evolution of the ion yields (Fig. 11(a)) and the 23rd harmonic intensity (Figs. 11(b) and 11(c)) from $CO_2$ molecules. The results of theoretical calculations are also included and reasonable

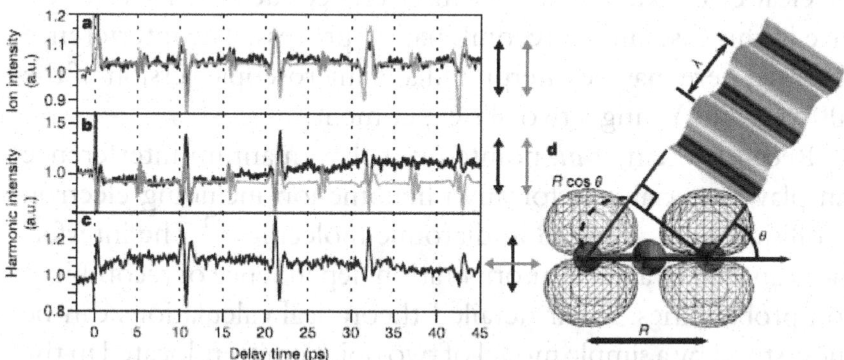


**Fig. 11.** The temporal evolution of the ion yield and the 23rd harmonic intensity from $CO_2$ molecules, and an illustration of the model of two point emitters. (a–c) The ion yield dominated by $CO_2^+$ (a) and the 23rd harmonics from $CO_2$ molecules (b, c) as a function of the pump–probe delay. The experimental conditions are identical with those of Fig. 9. The rotational period $T_{rot}$ of a $CO_2$ molecule is 42.7 ps. The results of theoretical calculations are also shown by gray curves. The rotational temperature of the $CO_2$ molecules is assumed to be 40 K and the highest degree of alignment $\langle \cos^2 \theta \rangle$ is estimated to be 0.70 at $3T_{rot}/4$. (d) A $CO_2$ molecule can be regarded as an elongated diatomic molecule. Two point emitters are located in two O nuclei. $\lambda$ is the de Broglie wavelength of an electron. $\theta$ is the orientation angle, i.e., the angle between the molecular axis and the polarization of the probe pulse. $R$ and $R \cos \theta$ are the distance between two O atoms and its projection, respectively. (Reprinted with permission from Ref. 20. Copyright (2005) by the Nature Publishing Group.)

agreement between the experiments and the calculations can be seen again. Since $T_{rot} = 42.7$ ps for linear $CO_2$ molecules, both of the ion and harmonic intensities modulate at every $T_{rot}/8$ as in the case of $O_2$. As in the cases of $N_2$ and $O_2$, the harmonic intensity in the parallel case (Fig. 11(b)) and that in the perpendicular case (Fig. 11(c)) shows the inverted modulations. However, when the polarizations are parallel to each other, the ion yield (Fig. 11(a)) and the harmonic intensity (Fig. 11(b)) also shows the inverted modulations as opposed to the cases of $N_2$ and $O_2$. The observations shown in Figs. 11(a) and 11(b) indicate that HHG is promoted when ionization is suppressed. This means that the recombination probabilities at a certain orientation angle overpower the ionization probabilities in the HHG process. The modulation inversions observed in Figs. 11(a) and 11(b) are the clear evidence of quantum interference (destructive interference in this case) in the recombination process. The interference observed here may be interpreted as a microscopic version of the 200-year-old Young's two-slit experiment.

Recently, Lein *et al.* pointed out that quantum interference can play a crucial role for any phenomenon including electron recollision such as HHG in diatomic molecules.[29] The interference can lead to a peculiar orientation dependence of recombination probabilities. Their detailed theoretical calculations can be understood by a simple model of two point emitters located in the positions of the nuclei. As mentioned above, since the HOMO of a $CO_2$ molecule is dominated by the two $p$ orbitals of the two O atoms, a $CO_2$ molecule can be regarded as an elongated diatomic molecule where the two point emitters are situated at the two O nuclei. The model of two point emitters applied here is illustrated in Fig. 11(d). The interference takes place within a single molecule and within one optical cycle. Therefore, simultaneous observations of both ion yields and harmonic signals can serve as a new route to probe instantaneous structures of molecular systems[30] as proposed in our original paper.[20] We further investigate ellipticity dependence of HHG from aligned $N_2$, $O_2$, and $CO_2$ molecules. It has been also found that the destructive interference in the recombination process affects the ellipticity dependence.[31]

## 5. Summary and Outlook

Following the introduction and the outline of the principle of molecular orientation, it has been shown how to achieve and observe one- and three-dimensional molecular orientation in Secs. 3.1 and 3.2, respectively. The efficacy of aligned molecules has been demonstrated by presenting two representative works from our laboratory in Sec. 4.

In this section, some future subjects to be challenged are presented. Concerning molecular alignment, it is quite important to increase the degree of alignment in the nonadiabatic regime. To do this, it could be a direct and powerful approach to optimally shape femtosecond pump pulses by using the degree of alignment evaluated from two-dimensional ion images as a feedback signal. We recently developed a technique with which two-dimensional ion images can be used as feedback signals for the first time. In fact, the optimal control of nonadiabatic molecular alignment is included as one of research subjects.

It is also important to increase the degree of orientation in the adiabatic regime. For this purpose, it should be effective to increase the electrostatic field, which is still much lower than that used in the brute force orientation. On the other hand, in precise spectroscopic measurements and experiments including the observation of photoelectrons, it is desirable to prepare a sample of oriented molecules in the (laser-)field-free condition. Noting that molecules can be oriented with combined electrostatic and intense, nonresonant laser fields as presented in Secs. 3.1 and 3.2, it may be a promising approach to use a shaped pulse which has a relatively long rising edge compared to the rotational period of the molecule and is suddenly switched off at the peak of the pulse. Such a suddenly switched-off pulse can be shaped with the plasma shutter technique.[32] Thereby, in the laser-field-free condition after the shaped pulse, it is expected that the same degree of orientation as that could be adiabatically achieved at the peak of the pulse is revived at the rotational period of the molecule. The feasibility study of this approach is under progress in our laboratory and its experimental realization is also one of our research subjects.

One of the most important challenges is to generate high-order harmonics from oriented molecules. With a sample of oriented molecules, we can expect the generation of even-order harmonics because of the breaking of inversion symmetry.

Finally, we really hope that many new and interesting phenomena will be found and investigated using a sample of aligned and/or oriented molecules in step with the progress of molecular manipulation techniques based on laser technologies.

**Acknowledgments**

The studies described in Secs. 3.1, 3.2, 4.1, and 4.2 are collaborations with S. Minemoto, H. Nanjo, H. Tanji, T. Suzuki, and T. Kanai. Their contributions are greatly appreciated. Those studies were supported by the Grants-in-Aid (No. 14102012 and No. 19204041) from the Japan Society for the Promotion of Science (JSPS) and that (No. 14077206) from the Ministry of Education, Culture, Sports, Science and Technology (MEXT) of Japan.

**References**

1. Special issue on *Stereodynamics of Chemical Reaction* [J. Phys. Chem. A **101**, 7461 (1997)].
2. P. B. Corkum, *Phys. Rev. Lett.* **71**, 1994 (1993).
3. D. Herschbach, *Eur. Phys. J. D* **38**, 3 (2006).
4. B. Friedrich and D. Herschbach, *Phys. Rev. Lett.* **74**, 4623 (1995).
5. B. Friedrich and D. Herschbach, *J. Phys. Chem.* **99**, 15686 (1995).

6. H. Sakai, C. P. Safvan, J. J. Larsen, K. M. Hilligsфe, K. Hald and H. Stapelfeldt, *J. Chem. Phys.* **110**, 10235 (1999).
7. J. J. Larsen, H. Sakai, C. P. Safvan, I. Wendt-Larsen and H. Stapelfeldt, *J. Chem. Phys.* **111**, 7774 (1999).
8. J. J. Larsen, K. Hald, N. Bjerre, H. Stapelfeldt and T. Seideman, *Phys. Rev. Lett.* **85**, 2470 (2000).
9. F. Rosca-Pruna and M. J. J. Vrakking, *Phys. Rev. Lett.* **87**, 153902 (2001).
10. K. F. Lee, D. M. Villeneuve, P. B. Corkum, A. Stolow and J. G. Underwood, *Phys. Rev. Lett.* **97**, 173001 (2006).
11. H. Stapelfeldt and T. Seideman, *Rev. Mod. Phys.* **75**, 543 (2003) and references therein.
12. T. Seideman and E. Hamilton, *Adv. At. Mol. Opt. Phys.* **52**, 289 (2006) and references therein.
13. B. Friedrich and D. Herschbach, *J. Chem. Phys.* **111**, 6157 (1999).
14. B. Friedrich and D. Herschbach, *J. Phys. Chem. A* **103**, 10280 (1999).
15. H. Sakai, S. Minemoto, H. Nanjo, H. Tanji and T. Suzuki, *Phys. Rev. Lett.* **90**, 083001 (2003).
16. S. Minemoto, H. Nanjo, H. Tanji, T. Suzuki and H. Sakai, *J. Chem. Phys.* **118**, 4052 (2003).
17. H. Sakai, S. Minemoto, H. Nanjo, H. Tanji and T. Suzuki, *Eur. Phys. J. D* **26**, 33 (2003).
18. H. Tanji, S. Minemoto and H. Sakai, *Phys. Rev. A* **72**, 063401 (2005).
19. T. Suzuki, S. Minemoto, T. Kanai and H. Sakai, *Phys. Rev. Lett.* **92**, 133005 (2004).
20. T. Kanai, S. Minemoto and H. Sakai, *Nature (London)* **435**, 470 (2005).
21. J. J. Larsen, I. Wendt-Larsen and H. Stapelfeldt, *Phys. Rev. Lett.* **83**, 1123 (1999).
22. J. Itatani, J. Levesque, D. Zeidler, H. Niikura, H. Pepin, J. C. Kieffer, P. B. Corkum and D. M. Villeneuve, *Nature (London)* **432**, 867 (2004).
23. M. Fukazawa, T. Inoue, S. Minemoto and H. Sakai, unpublished data.
24. Y. Silberberg, *Nature (London)* **430**, 624 (2004).
25. W. J. Walecki, D. N. Fittinghoff, A. L. Smirl and R. Trebino, *Opt. Lett.* **22**, 81 (1997).
26. T. Kanai, S. Minemoto and H. Sakai, in *Ultrafast Phenomena XIV*, eds. T. Kobayashi, T. Okada, T. Kobayashi, K. A. Nelson and S. De Silvestri (Springer, 2005), pp. 310–312.
27. X. M. Tong, Z. X. Zhao and C. D. Lin, *Phys. Rev. A* **66**, 033402 (2002).
28. T. Suzuki, S. Minemoto and H. Sakai, *Appl. Opt.* **43**, 6047 (2004).
29. M. Lein, N. Hay, R. Velotta, J. P. Marangos and P. L. Knight, *Phys. Rev. A* **66**, 023805 (2002).
30. J. P. Marangos, *Nature (London)* **435**, 435 (2005).
31. T. Kanai, S. Minemoto and H. Sakai, *Phys. Rev. Lett.* **98**, 053002 (2007).
32. B. J. Sussman, J. G. Underwood, R. Lausten, M. Yu. Ivanov and A. Stolow, *Phys. Rev. A* **73**, 053403 (2006).

## Chapter 4

# Sum Frequency Generation: An Introduction with Recent Developments and Current Issues

Mary Jane Shultz

*Pearson Research Laboratory*
*Chemistry Department Tufts University, Medford, MA 02155, USA*
*Mary.Shultz@Tufts.edu*

Nonlinear spectroscopy, both second harmonic and sum frequency generation — SHG and SFG — have proven to be powerful techniques for probing a variety of interfaces from the very dynamic, high vapor pressure liquid–air surface to buried interfaces between hydrophobic and hydrophilic phases to irregular and amorphous solid surfaces. With the advent of off-the-shelf laser systems, it has become easier and easier to collect nonlinear spectra. The major impediments to wide spread usage of nonlinear spectroscopy are the challenges in interpretation of the spectra produced. This work begins with an introduction to nonlinear spectroscopy based on an optical-geometrical view of the interaction between the probe beams and molecules in the interfacial region. The introduction serves as a basis for exploration of recent developments and current issues. Two case studies are included: examination of ions at the aqueous interface including evidence for $H_3O^+$ at the interface and investigation of molecular interactions on nonmetallic, nanostructured interfaces.

## 1. Introduction

Although issues still remain, it can be said that gas-phase reactions, interactions in solutions, and solid-phase processes are all reasonably understood. The same cannot be said of surfaces: They remain a major challenge for chemists and physicists, the last frontier. The reason that surfaces are so challenging is partly one of numbers: even in the gas phase, there are orders of magnitude more molecules in the probed volume than there are on a surface. A full monolayer corresponds to on the order of only $10^{14}$ molecules/cm$^2$ compared with a typical gas which has $10^{18}$–$10^{20}$ molecules/cm$^3$. At the same time, the molecular scale is short compared with the length scale of most probing techniques. Techniques that have a characteristic length scale comparable to or smaller than molecular dimensions generally require the target to be a regular array to produce a scattering pattern. Thus, it would seem an impossible challenge to probe irregular or dynamic interfaces such as the air–water interface, buried hydrophobic–hydrophilic interfaces that dominate biological processes, or even regular arrays like that of ice that are extremely dynamic at the most relevant temperatures. Finally, interfaces are most generally bordered by bulk phases; any signal from the bulk phase can easily overwhelm that from the interface. Ultra-high vacuum (UHV) methods overcome the bulk phase issue by evacuating the gas phase to extremely low number densities, restricting investigation either to low vapor pressure solids or to higher vapor pressure substrates that are briefly introduced into a chamber with high pumping efficiency.

If the system of interest does not meet UHV requirements and it is too dynamic or irregular for scattering techniques, how can it be investigated? Fortunately, optical techniques can still be used. Ideally, the optical technique would generate no signal from the bulk. Some interfaces nicely accommodate this requirement by partitioning the analyte to the surface. For example, nonvolatile surfactants in aqueous solution are biased to be on the surface. The length issue remains a challenge. Even short wavelength optical probes use light with a wavelength on the order of hundreds of nanometers — orders of magnitude larger than the hundreds of picometer scale of many molecules. The nonlinear optical techniques that solve the bulk interference and wavelength probe issue, the subject of this review, do so by using the change of electron density or electron polarizability that occurs right at the interface. (This same phenomenon gives rise to the apparent break in the handle of a spoon placed in a glass of water; the change in index of refraction gives the illusion of a break.) The two techniques are known as second harmonic generation (SHG) and sum frequency generation (SFG). This review is intended as a tutorial for those researchers entering this exciting field and is written in two major sections. The first presents a geometrical–optical view of the nonlinear process. The second contains a summary of current developments as well as descriptions of two results: the quest to determine the location of $H_3O^+$ in aqueous solutions (the evidence suggest that the distribution is biased toward the very top monolayer of the solution) and the challenge of probing the interface of a thin film consisting of nanometer-scale, nonmetallic particles.

A number of reviews of SHG and SFG have been written in the last several years,[1–15] so the question naturally arises, why another? The answer is that, to our knowledge, none of these reviews presents a geometrical–optical tutorial beginning with the incoming light beams through to the detected intensity. The advantage of the geometrical–optical view is that it can form the basis of a physical picture of the nonlinear process that is not simplified to the point of losing essential phenomena.

Both nonlinear techniques have seen increased activity in the last few years due to development of commercial picosecond and femtosecond lasers; the short-pulse lasers provide the high peak powers required for nonlinear spectroscopy. SHG accesses electronic resonances in the molecule requiring simultaneous adsorption of two photons, which is greatly facilitated by the high peak powers. SFG, as most often implemented involves an infrared and a visible photon. SFG benefits from high peak powers in two ways: generating the infrared and probing the surface. The nonlinear crystals used in the infrared generation have become increasingly reliable. As a result, it is easier and easier to collect nonlinear spectra. These spectra are rich in information. With the information richness comes a challenge: deconvoluting the wealth of orientation and polarization information in the spectra. The tutorial section of this review is intended to lower the activation barrier to deriving meaningful data from the nonlinear spectra.

## 2. Electric Fields and Orientation Factors

Both nonlinear techniques begin with light impinging on the surface. Light impinging on a medium influences the medium via the electric and magnetic fields of the light. In the electric dipole approximation, only the electric field contributes to the response. The response of the medium is the result of polarization of the electrons of the medium by the incoming electric field. Since the light electric field is oscillating, the resultant polarization, $\mathcal{P}$, also oscillates. The oscillating electron polarization subsequently radiates giving rise to the output light. To first order, the output is merely proportional to the input light fields; to second order it is quadratic in the fields:

$$\mathcal{P} = \tilde{\tilde{\alpha}}^{(1)}\tilde{E}_1 + \tilde{\tilde{\alpha}}'^{(1)}\tilde{E}_2 + \tilde{\tilde{\tilde{\chi}}}^{(2)} : \tilde{E}_1\tilde{E}_2 + \cdots, \tag{1}$$

where $\tilde{\tilde{\alpha}}^{(1)}$ ($\tilde{\tilde{\alpha}}'^{(1)}$) is a matrix describing the first order response to the incoming field, $\tilde{E}_1(\tilde{E}_2)$. The second order response, $\tilde{\tilde{\tilde{\chi}}}^{(2)}$, is a tensor that gives rise to SHG, SFG, and difference frequency generation (DFG) depending on the relationship between the input and output frequencies. All of these are generated at the sample, however this review focuses on SHG and SFG.

The goal of surface spectroscopy is to unravel the relationship between the direction of the electric fields in the incoming beams, $\tilde{E}_1$ and $\tilde{E}_2$, and the observed intensity of the nonlinear beam, which is related to the square of the polarization, $\mathcal{P}$. The electric

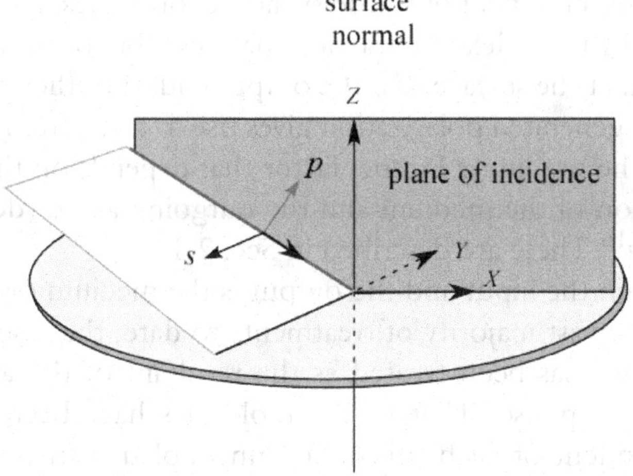


**Fig. 1.** The plane of incidence is defined by the surface normal and the propagation direction. Polarization is defined with respect to this plane: *p*-polarized in the plane and *s*-polarized perpendicular to the plane.

field direction is commonly referenced to the input–output plane; this plane is defined by the beam propagation direction and the surface normal (Fig. 1). An electric field in the plane is termed *p*-polarized; a field perpendicular to the plane is *s*-polarized. Note that the field of *s*-polarized light is along the *Y* axis at the surface. In contrast, the field of *p*-polarized light at the surface has a component that is along the *X* axis and a component that is along the *Z* axis; the balance between *X* and *Z* depends on the angle between the propagation direction and the surface normal. This distinction between *p*- and *s*-polarized light is particularly important when analyzing the *p*-polarized output generated from two *p*-polarized input beams. This is elaborated around Eq. (9).

Since the surface is the focus, all components are referenced to the surface. Briefly, the influence of the incoming radiation on the surface is described by the efficiency with which the incoming radiation couples into the surface — this is described by the Fresnel factors (Sec. 2.1). These Fresnel factors consist of a combination of a geometrical term describing the projection of the incoming electric field onto the surface coordinates, and a term governed by the index of refraction that describes bending of the light beam at the surface. On the output end, the efficiency with which the generated polarization gives rise to an emitted beam is given by the nonlinear Fresnel factor that depends on the index of refraction of the medium and the outgoing angle (described below).[12,16] These are described in Sec. 2.1.

Between the input and the output is the medium of the surface. In the vast majority of treatments to date, the response of the medium has been treated as the resultant of the averaged molecular response. That is, the molecules have been treated as independent of each other, i.e., uncoupled with no collective modes. Particularly for the solvent in a strongly coupled system, such an independent molecule approach may not be valid and the system should more properly be treated as a large molecule with collective modes.[17,18] Although collective modes strain computational resources, the relationship between the molecular response and that of the surface is similar to the independent molecule approach outlined here. Within the independent molecule approach, determining the response of the medium consists of determining the projection of the molecular response — either the electronic transition moment (SHG) or the vibrational dipole times the Raman polarizability (SFG) — onto the surface fields (Sec. 2.2). That is (Table 1 and Fig. 2) the molecular coordinate system $(a, b, c)$ is projected onto the surface coordinate system $(x, y, z)$.

Projecting the molecular coordinate system $(a, b, c)$ onto the surface system $(x, y, z)$ can be described by a sequence of

**Table 1.** Notation for coordinate frames.

| System | Laboratory | Surface | Molecular | Normal mode |
|---|---|---|---|---|
| Coordinate | $XYZ$ | $xyz$ | $abc$ | $ABC$ |
| Hyperpolarizability | $\chi_{IJK}$ | $\chi_{ijk}$ | $\beta_{\alpha\beta\gamma}$ | |

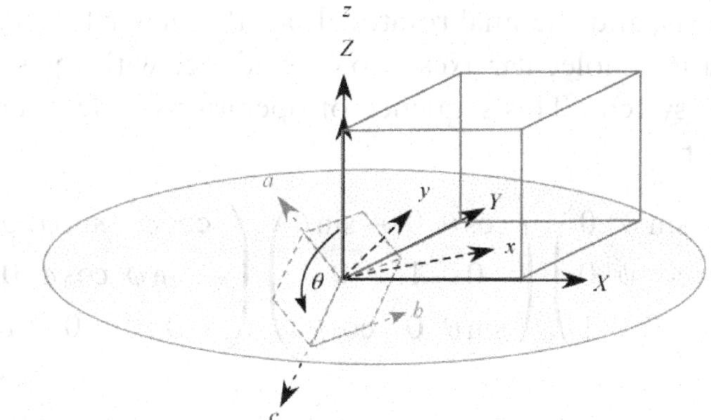


**Fig. 2.** Illustration of the relationship among the laboratory $(X, Y, Z)$, the surface $(x, y, z)$, and the molecular $(a, b, c)$ coordinate systems. If the surface is isotropic, as in a liquid surface, then the laboratory and surface coordinate systems coincide. Often, the important relationship is the tilt of the molecular coordinate system relative to the surface normal $(z)$, indicated by the angle $\theta$.

**Table 2.** Angle glossary — Angles defined with respect to the surface coordinate system.

| Angle | Measures |
|---|---|
| $\theta$ | Tilt angle between surface $z$ and molecular $c$ axis |
| $\phi$ | Twist angle between the molecular $a$, $c$ plane and surface $x$, $y$ plane |
| $\psi$ | Rotation of the molecular $a$ with respect to the surface $x$ axis (free rotation in isotropic surface) |
| $\eta$ | Between surface $z$ and input beam propagation direction |
| $\delta$ | Between visible and infrared input planes |
| $\Theta$ | Polarization: angle between generated light electric field, $\tilde{E}_{SF}$, and output plane |

rotations that bring the molecular axes into coincidence with those of the surface. This is known as an Euler angle projection and is the product of three rotational matrices (see Table 2 for angle definitions). Starting with the molecular frame attached to the molecule, the first operation rotates about the molecular $c$ (highest symmetry) axis by an angle $\phi$ to bring the molecular $b$ axis to the surface plane, the second rotates about the new $b$ axis by an angle $\theta$ so that the molecular $c$ axis coincides with the surface $z$ axis, and the final rotates about the new $c$ by an angle $\psi$ to bring the molecular axes into coincidence with the surface coordinate system. This sequence of operations is described by the product

$$\begin{pmatrix} \cos\psi & \sin\psi & 0 \\ -\sin\psi & \cos\psi & 0 \\ 0 & 0 & 1 \end{pmatrix} \begin{pmatrix} \cos\theta & 0 & -\sin\theta \\ 0 & 1 & 0 \\ \sin\theta & 0 & \cos\theta \end{pmatrix} \begin{pmatrix} \cos\phi & \sin\phi & 0 \\ -\sin\phi & \cos\phi & 0 \\ 0 & 0 & 1 \end{pmatrix}. \tag{2}$$

The product is a $3\times3$ matrix, hence contains nine elements — three describe the projection of the molecular $a$ axis onto the surface $x$, $y$, and $z$ axes, three more describe projection of the $b$ axis, and the final three correspond to projection of the $c$ axis. Since there are three surface fields: the two input beams and the output polarization, there are three $3 \times 3$ matrices that describe the projection of the molecular axes onto the surface fields. A stack of three, $3 \times 3$ matrices is a $3 \times 3 \times 3$ tensor containing a total of 27 elements. Labeling the elements of each $3 \times 3$ matrix with the row and column as (row, column), enables reading the projection of the $\alpha\beta\gamma$ molecular axes onto the $ijk$ components of the surface fields from the product matrix in Eq. (2). The first projection is that of the molecule onto the surface SF polarization: it is the $(i, \alpha)$ element of the first $3\times3$. Multiplying the projection of the molecule onto the SF polarization times the projection of the molecule onto the surface visible field $(j, \beta)$ and further multiplying times the projection of the molecule onto the infrared field $(k, \gamma)$ gives the projection of the molecular polarization, $\beta_{\alpha\beta\gamma}$ onto the surface polarization $\chi_{ijk}$. (The interested reader can find these 27 elements written out in Hirose *et al.*[12]) Each of these 27 elements is averaged over all equivalent molecular configurations. For a surface with an isotropic surface plane, the average is obtained by integrating each tensor element over the cone of equivalent configurations, i.e., integrate

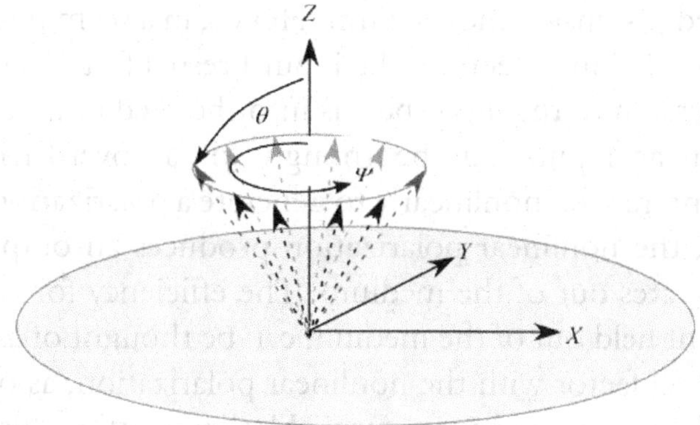


**Fig. 3.** For a rotationally isotropic surface, all locations on the same tilt cone are equivalent. The field projections are thus integrated over this cone.

$0° \leq \psi \leq 360°$ (Fig. 3). The result is an average orientation tensor element denoted $\langle ijk|\alpha\beta\gamma\rangle$.

Due to the averaging over equivalent orientations, many tensor elements are equal to zero. This projection and orientation averaging are discussed in greater detail in Sec. 2.2.

## 2.1. *Fresnel factors and propagation direction*

The efficiency with which the incoming radiation produces a field in the surface is described by the linear Fresnel factors; the same factors that describe reflection and refraction of light beams at an interface. These factors describe the amplitude of the electric field, $E$, at various locations relative to an incident field, hence describe the effect of the input infrared and visible fields on the interface. Interferences between the resultant fields in the interface give rise to the harmonics and sum and difference frequencies, hence the importance of the Fresnel factors. These factors differ for SHG or SFG in reflection compared to that in transmission. Since most experimental arrangements collect the nonlinear beam in reflection, this case is described here. The interference occurs just inside the medium: that is, the nonlinear polarization is generated just inside the medium. Hence, in a seeming contradiction, for SFG in reflection, the input Fresnel factors are those for transmission of the input beams into the medium.

The linear inputs can be thought of as overdriving the medium, interacting nonlinearly to generate a polarization. Once generated, the nonlinear polarization produces an output field that propagates out of the medium. The efficiency for coupling the resultant field out of the medium can be thought of as a nonlinear Fresnel factor with the nonlinear polarization, as opposed to a light beam, as the source term. Mathematically, this source term is pure imaginary.

Technically, both the input and the output Fresnel factors are calculated from Maxwell's equations and are determined by continuity of the tangential components of the $\tilde{E}$ and $\tilde{B}\mu^{-1}$ fields across the interface.[16] These factors are listed in Table 3 for the case of SFG.[1,16,19,20] (The factors for SHG can be determined from those of SFG by setting $\omega_1 = \omega_2$.)

The $K_X$ and $K_Z$ Fresnel factors derive from the parallel transmission factors as follows. The $K_X$ factor is the parallel transmission amplitude projected onto the surface $X$ axis via the term cos

**Table 3.** Linear ($K$) and nonlinear ($L$) optical factors.[1,16,19,20,a]

| | $L$(SF) | $K$ |
|---|---|---|
| $X$ | $\frac{-2n_{1,\text{SF}}(\omega)\cos\eta_{t,\text{SF}}}{n_{1,\text{SF}}\cos\eta_{t,\text{SF}}+n_{2,\text{SF}}\cos\eta_{r,\text{SF}}}$ | $\frac{2n_{1,\omega}\cos\eta_{i,\omega}\cos\eta_{t,\omega}}{n_{1,\omega}\cos\eta_{t,\omega}+n_{2,\omega}\cos\eta_{i,\omega}}$ |
| $Y$ | $\frac{2n_{1,\text{SF}}\cos\eta_{r,\text{SF}}}{n_{1,\text{SF}}\cos\eta_{r,\text{SF}}+n_{2,\text{SF}}\cos\eta_{t,\text{SF}}}$ | $\frac{2n_{1,\omega}\cos\eta_{i,\omega}}{n_{1,\omega}\cos\eta_{i,\omega}+n_{2,\omega}\cos\eta_{t,\omega}}$ |
| $Z$ | $\frac{2n_{1,\text{SF}}^2 n_{2,\text{SF}}\cos\eta_{r,\text{SF}}}{(n'_{\text{SF}})^2[n_{1,\text{SF}}\cos\eta_{t,\text{SF}}+n_{2,\text{SF}}\cos\eta_{r,\text{SF}}]}$ | $\frac{2n_{1,\omega}n_{2,\omega}^2\cos\eta_{i,\omega}\sin\eta_{t,\omega}}{(n'_{\omega})^2[n_{1,\omega}\cos\eta_{t,\omega}+n_{2,\omega}\cos\eta_{i,\omega}]}$ |

[a] $XZ$ = plane of incidence, $XY$ = surface plane, $\eta_i$ = incident angle, $\eta_t$ = transmitted angle, $\eta_r$ = reflectance angle, $n_{2,\omega}$ = refractive index of medium 2 at frequency $\omega$, $n'$ = refractive index of the interfacial layer (in recent work, $n'$ is assigned a value midway between that of the two bounding media). $K_Z$ differ from that of Shen and coworkers[20] due to application of Schnell's law to assist with connection with Fresnel factors in Born and Wolf.[19]

$\eta_{t,\omega}$. The $K_Z$ factor is similarly projected via $\sin \eta_{t,\omega}$ and is further modified by the factor $(n_2/n')^2$ due to treatment of the interfacial layer as a thin slab between the two bounding media. The index for the interfacial layer, $n'$, is discussed further in Sec. 4.1. In recent work, $n'$ has been assigned a value midway between the two bounding media.

Note that the above-mentioned Fresnel factors assume that medium 2 is isotropic being described by a single index of refraction. Extension to the case of a substrate that is birefringent is straightforward, though more complex. Birefringent substrates have two indices of refraction. The simpler cases consist of the optic axis (the *c* axis) either in the input plane or perpendicular to it. If the optic axis is in the input plane and parallel to the surface the refractive index for the *y* and *z* components differ from that for *x*. If the optic axis is in the input plane but perpendicular to the surface, the refractive index for the *x* and *y* components differ from that for *z*. If the optic axis is perpendicular to the input plane, then it must be parallel to the surface and the refractive index for the *y* component differs from that for *x* and *z*.

In the language of birefringent crystals, the component of the electric field that is perpendicular to the optic axis is the ordinary or *o*-ray; the component along the optic axis is the extraordinary or *e*-ray.[21] If the optic axis is parallel to the surface and in the plane of incidence, the *x* component of *p* polarized light is an *e*-ray and the *z* component is an *o*-ray; the balance between the *x* and *z* components of the electric field changes as the electric field interacts with the surface. If the optic axis is perpendicular to the surface but in the input plane, the *x* component is an *o*-ray and the *z* component is an *e*-ray. Once again the balance changes as the field interacts with the surface. When the optic axis is in the input plane, *s*-polarized light (the *y* component) is an *o*-ray whether the optic axis is parallel or perpendicular to the surface. If the optic axis is parallel to the surface and perpendicular to the input plane, then both the *x* and *z* components are *o*-rays and

$s$-polarized light (the $y$ component) is an $e$-ray. Finally, if the optic axis is at an angle to the surface, then one gets a combination of these cases.

As can be seen in Table 3, the Fresnel factors require knowledge of the direction of propagation of the sum frequency (or second harmonic) wave. This direction is determined by momentum matching at the interface

$$n_{1,\mathrm{SF}}^2\omega_{\mathrm{SF}}^2 \sin^2 \eta_{\mathrm{r,SF}} = n_{1,\omega 1}^2\omega_1^2 \sin^2 \eta_{\mathrm{i},1} + n_{1,\omega 2}^2\omega_2^2 \sin^2 \eta_{\mathrm{i},2} \pm 2n_{1,\omega 1}n_{1,\omega 2}\omega_1\omega_2 \sin \eta_{\mathrm{i},1} \sin \eta_{\mathrm{i},2} \cos \delta, \quad (3)$$

where SF, 1, and 2 refer to the sum frequency and input beams 1 and 2 respectively, $n_{1,\omega i}$ is the index of refraction of the $i$th beam in the input medium. The input angle $\eta_i$ is measured from the surface normal as is the angle of the reflected SF beam, $\eta_{\mathrm{r,SF}}$. The angle $\delta$ is the angle between input planes 1 and 2. In most applications, $\delta = 0$ but there are special experimental considerations in which it is convenient to have a nonzero $\delta$. If $\delta = 0$, then Eq. (3) simplifies to

$$n_{1,\mathrm{SF}}\omega_{\mathrm{SF}} \sin \eta_{\mathrm{r,SF}} = n_{1,\omega 1}\omega_1 \sin \eta_{i,1} \pm n_{1,\omega 2}\omega_2 \sin \eta_{i,2}. \quad (4)$$

The plus sign of Eqs. (3) and (4) applies if the input beams are copropagating and the minus sign if they are counterpropagating.

### 2.2. *Orientation factors*

In many interfacial studies, one of the goals is to determine the orientation of the interfacial molecules. This information is contained in the relationship between the surface hyperpolarizability — the nonlinear or higher order polarization response of the medium — and that of the molecules constituting the interfacial region. (Since the output intensity is affected by the molecular orientation, surface concentration measurements also require knowledge of the orientation.) Within the independent molecule approximation, the surface hyperpolarizability is the orientation average of the molecular hyperpolarizabilities. It is the orientation averaging constrained by the symmetry of the surface that results in many elements of the surface hyperpolarizability tensor averaging to zero. Similarly, molecular symmetry often simplifies the molecular hyperpolarizability tensor. Resonances, either electronic (SHG) or vibrational (SFG) further limit the number of significant tensor elements. The nonzero, nonresonant tensor elements may be combined with the nonresonant background from the substrate or solvent.

The nonresonant background is a property of the system and not under the experimenter's control. In contrast, quadrupole and higher order response terms result from driving the medium beyond a quadratic response. Often the higher order response terms can be diminished by operating within an optimum input intensity: high enough to generate a signal but low enough that the output intensity is quadratic in the input intensity (SHG) or linear in both the infrared and visible input intensities (SFG).

The following discussion of orientation factors is based on an independent-molecule, dipole approximation. (As indicated above, in the case where collective modes are important, treating the surface region as one large molecule is a straightforward extension of this analysis once the surface modes have been determined.) To simplify the discussion, from this point forward, only vibrational sum frequency generation is treated.

Progress in determining the orientation factors consists of (a) determining the absolute orientation via interference with a reference system,[22–24] (b) narrowing the orientation by exploiting orthogonal modes of the surface molecule,[25,26] (c) using a null angle method to more accurately determine the orientation,[7,27,28] and (d) treating the system as a large molecule in an asymmetric environment for the solvent or collective modes.[18] Each of these is discussed below in Secs. 3.1–3.3 and 4.2 following a general discussion of the orientation averaging.

### 2.2.1. *Simplification of the orientation tensor*

The orientation average consists of determining the projection of the product of the molecular infrared and Raman transition moments onto the surface infrared and visible oscillating electric fields, then averaging over all allowed molecular orientations. The infrared (Raman) transition moment is described within the molecular coordinate system, so projection onto the surface coordinate system consists of projecting the molecular $(a, b, c)$ axes onto the surface $(x, y, z)$ axes[12] and averaging over all molecular orientations. As discussed above, this is the origin of the $3 \times 3 \times 3$ orientation tensor.

Surface symmetry simplifies this tensor. For example, if the surface contains a $y$–$z$ reflection plane, then the surface hyperpolarizability must be invariant to interchange of $+x$ and $-x$. Since the projection onto $+x$ is equal to but opposite in sign to the projection onto $-x$, $\chi_{ijx} = \chi_{ij-x} = -\chi_{ijx}$ $(i, j \neq x)$. The only quantity that can be equal to minus itself is zero, so $\chi_{ijx} = 0$. Similarly $\chi_{ixj} = -\chi_{ixj} = 0$ and $\chi_{xij} = -\chi_{xij} = 0$. Hence, of the 27 possible surface hyperpolarizability elements, all with either one or three $x$ indices are zero. Likewise, for any surface with an $x$–$z$ reflection plane, all tensor elements with one or three $y$ subscripts are zero. This reduces the number of nonzero surface tensor elements to seven: $\chi_{xxz}$, $\chi_{xzx}$, $\chi_{zxx}$, $\chi_{yyz}$ $\chi_{yzy}$, $\chi_{zyy}$, and $\chi_{zzz}$.

For vibrationally resonant, visible nonresonant SFG, the first two subscripts reference the Raman transition moment and the last the infrared transition element, so $\chi_{xzx} = \chi_{zxx}$ and $\chi_{yzy} = \chi_{zyy}$.[a] In addition, if the surface is isotropic (e.g., a liquid surface), then $x$ and $y$ are equivalent, so $\chi_{xzx} = \chi_{yzy}$ leaving three, unique, nonzero elements: $\chi_{xxz}$, $\chi_{xzx}$, and $\chi_{zzz}$.

These macroscopic surface hyperpolarizability elements, $\chi_{ijk}$, result from the molecular hyperpolarizability elements, $\beta_{\alpha\beta\gamma}$, and

[a]Note: for SHG, the last two subscripts are interchangeable, since both input photons are the same.

are related via the averaged orientation tensor, the average of the projection of the $\alpha\beta\gamma$ onto the $ijk$, $\langle ijk \mid \alpha\beta\gamma\rangle$:

$$\chi_{ijk} = N_s \sum_{\alpha\beta\gamma} \langle ijk|\alpha\beta\gamma\rangle \beta_{\alpha\beta\gamma}, \tag{5}$$

where $N_s$ is the number of scatters. A general formulation for $\langle ijk \mid \alpha\beta\gamma\rangle$ has been given by Hirose and coworkers.[14] Examples are given in Sec. 2.3.1.

### 2.3. *Observed intensity*

The observed intensity is related to the surface hyperpolarizability and is polarization dependent. Three polarizations are required to describe the experiment: the polarization of the SF beam, the visible input, and the infrared input in that order. Thus, there are eight potential polarization combinations. If the surface has an $x$–$z$ mirror plane, then those combinations with an odd number of $s$ polarized beams have zero intensity. As with the surface hyperpolarizability, the SF and visible represent a Raman process, so $I_{sps}$ and $I_{pss}$ differ only by a Fresnel factor, thus there are three unique configurations: *ssp*, *sps*, and *ppp*.

The output intensity is related to the square of the second order polarization, $\chi^{(2)}$:[1]

$$I(\omega) = \frac{32\pi^3\omega^2 \sec \eta_{\mathrm{SF}}^2}{c^3 n_1(\omega_{\mathrm{SF}}) n_1(\omega_1) n_1(\omega_2)} \left| \tilde{E}(\omega_{\mathrm{SF}}) \cdot \tilde{\tilde{\tilde{\chi}}}^{(2)} : \tilde{E}(\omega_{\mathrm{vis}}) \tilde{E}(\omega_{\mathrm{IR}}) \right|^2, \tag{6}$$

where $\tilde{E}(\omega)$ is a vector describing the direction and magnitude of the input or output fields and $\tilde{\tilde{\tilde{\chi}}}^{(2)}$ is constituted from the component hyperpolarizabilities, $\chi_{IJK}$. The output intensity is thus proportional to the hyperpolarizabilities weighted with their respective Fresnel and optical factors, $K$ and $L$, and the input field strengths:

$$
\begin{aligned}
I_{ssp} &\propto |L_Y \chi_{YYZ} K_Y^{\text{vis}} K_Z^{\text{IR}} e_{\text{vis}}^{s} e_{\text{IR}}^{p}|^2,\\
I_{sps} &\propto |L_Y \chi_{YZY} K_Z^{\text{vis}} K_Y^{\text{IR}} e_{\text{vis}}^{p} e_{\text{IR}}^{s}|^2,\\
I_{ppp} &\propto |\cos\eta_{\text{r,SF}} L_X \chi_{XXZ} K_X^{\text{vis}} K_Z^{\text{IR}} e_{\text{vis}}^{p} e_{\text{IR}}^{p}\\
&\quad + \cos\eta_{\text{r,SF}} L_X \chi_{XZX} K_Z^{\text{vis}} K_X^{\text{IR}} e_{\text{vis}}^{p} e_{\text{IR}}^{p}\\
&\quad + \sin\eta_{\text{r,SF}} L_Z \chi_{ZXX} K_X^{\text{vis}} K_X^{\text{IR}} e_{\text{vis}}^{p} e_{\text{IR}}^{p}\\
&\quad + \sin\eta_{\text{r,SF}} L_Z \chi_{ZZZ} K_Z^{\text{vis}} K_Z^{\text{IR}} e_{\text{vis}}^{p} e_{\text{IR}}^{p}|^2.
\end{aligned}
\tag{7}
$$

(Details of the tensor product in (7) are contained in Appendix A.) In the following discussion, it is useful to define an amplitude for the emitted sum frequency that is equal to the term inside the absolute squares in Eq. (7) — this is the amplitude divided by the combination of factors $\{32\pi^2\omega^2 \sec \eta_{\text{SF}}^2/c^3 n_1(\omega_{\text{SF}}) n_1(\omega_1) n_1(\omega_2)\}^{1/2}$:

$$
\begin{aligned}
A_{ssp} &\equiv L_Y \chi_{YYZ} K_Y^{\text{vis}} K_Z^{\text{IR}} e_{\text{vis}}^{s} e_{\text{IR}}^{p},\\
A_{sps} &\equiv L_Y \chi_{YZY} K_Z^{\text{vis}} K_Y^{\text{IR}} e_{\text{vis}}^{p} e_{\text{IR}}^{s},\\
A_{ppp} &\equiv \cos\eta_{\text{r,SF}} L_X \chi_{XXZ} K_X^{\text{vis}} K_Z^{\text{IR}} e_{\text{vis}}^{p} e_{\text{IR}}^{p}\\
&\quad + \cos\eta_{\text{r,SF}} L_X \chi_{XZX} K_Z^{\text{vis}} K_X^{\text{IR}} e_{\text{vis}}^{p} e_{\text{IR}}^{p}\\
&\quad + \sin\eta_{\text{r,SF}} L_Z \chi_{ZXX} K_X^{\text{vis}} K_X^{\text{IR}} e_{\text{vis}}^{p} e_{\text{IR}}^{p}\\
&\quad + \sin\eta_{\text{r,SF}} L_Z \chi_{ZZZ} K_Z^{\text{vis}} K_Z^{\text{IR}} e_{\text{vis}}^{p} e_{\text{IR}}^{p}.
\end{aligned}
\tag{8}
$$

Equations (7) and (8) optical factors are given in Table 3 and $e_{\text{vis}}$ ($e_{\text{IR}}$) is the magnitude of the visible (infrared) electric field with polarization specified by the superscript. Note in particular, that the *ppp* SF signal intensity is not simply the sum of the squares of the separate $X$ and $Z$ components, but rather the square of the sum. This distinction stems from the requirement for momentum conservation: The laboratory frame $X$ and $Z$ components of the emitted electric field are not independent, but linked by momentum conservation. Specifically, the $p$ polarized surface polarization is a combination of the laboratory $X$ and $Z$ polarized components:

$$|\mathcal{P}_p|^2 = |\mathcal{P}_X \cos\eta_{SF} + \mathcal{P}_Z \sin\eta_{SF}|^2. \tag{9}$$

In general, the vector sum of the $X$ and $Z$ components of the generated polarization is in the input plane but not perpendicular to the output propagation direction, which is determined by momentum conservation. Only that component of the generated polarization that is perpendicular to the output propagation direction can couple out of the interface. This component is determined by the projection indicated in Eq. (9). The sum, then square gives rise to interferences that are exploited in null angle spectroscopy (described in Sec. 3.3) and in nonlinear optical null ellipsometry (described in Sec. 4.1). (This important distinction was missed in a recent tutorial on SFG.)[29]

For a rotationally isotropic surface, the surface coordinate system $(x, y, z)$ coincides with the laboratory coordinate system $(X, Y, Z)$. If the surface is not rotationally invariant, the geometric relationship between the two systems must be inserted to convert Eqs. (7) and (8) from the laboratory frame to the surface frame. In most applications, the surface is rotationally invariant; resulting in a one-to-one relationship between surface and laboratory coordinates. This is the case that is treated from this point forward.

Note that, except for the phase (the sign), $I_{ssp}(I_{sps})$ measures $\chi_{yyz}(\chi_{yzy})$. Hence, if the phases of $\chi_{yyz}$ and $\chi_{yzy}$ could be determined, then measurement of $I_{ppp}$ would determine $\chi_{zzz}$ and the molecular orientation would be known. One method, described in Sec. 3.1, for determining the phases is to measure the intensity relative to a surface with a known phase.[22–24]

### 2.3.1. *Molecular examples*

The preceding discussion is general. It is useful to apply it to a specific system to make the relationships more transparent. Treatment of a specific molecular system is simplified using the molecular symmetry in an analogous manner to the surface hyperpolarizability. In addition, the significant molecular hyperpolarizability elements are selected by the infrared frequency. So, for example, for the symmetric stretch frequency of water, the resonant elements are $\beta_{aac}$ and $\beta_{ccc}$; a combination of the Raman transition moments, $\alpha_{aa}$ and $\alpha_{cc}$ with the infrared transition, $\mu_c$. For the antisymmetric stretch the resonant element is $\beta_{aca}$; a product of the Raman transition moment $\alpha_{ac}$ with the infrared transition, $\mu_c$. The relationship between the surface and molecular hyperpolarizability thus simplifies. For the symmetric stretch,

$$\begin{aligned}
\chi_{xxz} &= \beta_{aac}\cos\theta - (1/2)(\beta_{aac} - \beta_{ccc})(\cos\theta - \cos^3\theta) \\
&= 1/2(\beta_{aac} + \beta_{ccc})\cos\theta + 1/2(\beta_{aac} - \beta_{ccc})\cos^3\theta, \\
\chi_{xzx} &= 1/2(\beta_{ccc} - \beta_{aac})(\cos\theta - \cos^3\theta) \\
&= 1/2(\beta_{ccc} - \beta_{aac})\cos\theta - (1/2)(\beta_{ccc} - \beta_{aac})\cos^3\theta \\
\chi_{zzz} &= \beta_{ccc}\cos^3\theta + \beta_{aac}(\cos\theta - \cos^3\theta) \\
&= \beta_{aac}\cos\theta + (\beta_{ccc} - \beta_{aac})\cos^3\theta.
\end{aligned} \tag{10}$$

and for the antisymmetric stretch

$$\begin{aligned}
\chi_{xxz} &= -\beta_{caa}(\cos\theta - \cos^3\theta)\cos^2\phi \\
\chi_{xzx} &= (1/2)\beta_{caa}(1 - 2\sin^2\theta\cos^2\phi)\cos\theta \\
\chi_{zzz} &= 2\beta_{caa}(\cos\theta - \cos^3\theta)\cos^2\phi,
\end{aligned} \tag{11}$$

where $\theta$ is the tilt angle of the molecular symmetry axis with respect to the surface normal and $\phi$ is the twist angle of the molecular plane relative to the surface plane. Note that this analysis treats the water molecules as independent: collective modes and significant intermolecular vibrational coupling have been neglected. While this is probably a valid approximation for water on nonaqueous surfaces, hydrogen bonding makes it inadequate for aqueous solutions.[18]

The other frequently encountered molecular symmetry is that with a threefold rotation axis such as in $-CH_3$ or $NH_3$. For these, the relationships among the surface and molecular hyperpolarizabilities are the same as Eq. (10) for the symmetric stretch (both are one dimensional along the symmetry axis) and for the antisymmetric stretch:

$$\begin{aligned}
\chi_{xxz} &= -(1/2)\beta_{aaa}\sin^3\theta\cos 3\phi - \beta_{caa}(\cos\theta - \cos^3\theta),\\
\chi_{xzx} &= -(1/2)\beta_{aaa}\sin^3\theta\cos 3\phi + \beta_{caa}\cos^3\theta,\\
\chi_{zzz} &= \beta_{aaa}\sin^3\theta\cos 3\phi + 2\beta_{caa}(\cos\theta - \cos^3\theta).
\end{aligned} \tag{12}$$

When $\phi$ is random, often the case for $C_3$ molecules or groups, Eq. (12) is integrated over $0° \le \phi \le 360°$ and the first term vanishes for each $\chi$.

## 3. Recent Developments

### 3.1. *Absolute orientation determination with a reference*

The molecular hyperpolarizability is a function of the frequency, $\omega$, the infrared dipole, $\mu_c$, the Raman polarizability, $\alpha_{ab}$, and the bandwidth, $\Gamma_q$, of the $q$th vibrational resonance

$$\beta_{abc}(\omega, q) = \frac{\alpha_{ab}\mu_c}{\omega - \omega_q + i\Gamma_q}. \tag{13}$$

As indicated in Eq. (13), resonances have a finite band width, so for any frequency, $\omega$, the observed intensity is the result of multiple, overlapping resonances. In addition, there is also generally a nonresonant background, $\chi_{\text{NR}}$. All resonances and the nonresonant part must be included to determine the surface hyperpolarizability, which is the sum of all the contributions.

$$\chi_{ijk} = \sum_{q=1}^{n} \chi_{ijk}^{q} + \chi_{\text{NR}} = \sum_{q=1}^{n} \sum_{\alpha\beta\gamma} \langle ijk \mid \alpha\beta\gamma \rangle \beta_{\alpha\beta\gamma}(\omega, q) + \chi_{\text{NR}}, \tag{14}$$

where $\langle | \rangle$ denotes the projection of the molecular coordinates onto the surface frame. The observed intensity, Eq. (6), is proportional to the absolute square of the sum of the $\chi_{ijk}$. For example, from an isotropic surface containing a $C_{2v}$ molecule, where only the symmetric ($\chi_{yyz}^{s}$) and antisymmetric ($\chi_{yyz}^{a}$) stretches are within the observed bandwidth

$$\begin{aligned} |A_{ssp}|^2 &= |L_y K_y^{\text{vis}} K_Z^{\text{IR}} e_{\text{vis}}^{s} e_{\text{IR}}^{p} (\chi_{yyz}^{a} + \chi_{yyz}^{s}) + \chi_{\text{NR}}|^2 \\ &= |L_y K_y^{\text{vis}} K_Z^{\text{IR}} e_{\text{vis}}^{s} e_{\text{IR}}^{p} [\beta_{aac} \cos\theta - 1/2(\beta_{aac} - \beta_{ccc} \\ &\quad - \beta_{caa} \cos^2\phi)(\cos\theta - \cos^3\theta)] + \chi_{\text{NR}}|^2. \end{aligned} \tag{15}$$

It is this sum then square, as indicated in Eq. (15) that is responsible for the interferences that both complicates the interpretation of SFG spectra and enriches the information content including enabling determination of the absolute orientation. Absolute orientation cannot be deduced directly because the observed intensity is proportional to the square of the hyperpolarizability: A dipole with positive orientation (dipole pointing out of the bulk into the adjoining medium) produces the same observed intensity as one with a negative orientation (one pointing into the bulk). The absolute orientation information is lost because only the intensity is measured. The orientation information can be captured if the polarization mixes with another polarization of known phase, the two polarizations destructively

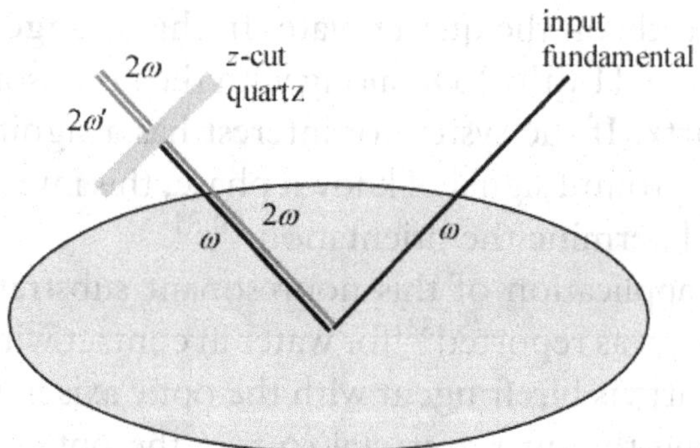


**Fig. 4.** Absolute orientation with SHG: the reflected fundamental, $\omega$, and second harmonic, $2\omega$, are directed into a $z$-cut quartz, which generates an additional second harmonic, $2\omega'$, of known phase. The two second harmonic beams, $2\omega$ and $2\omega'$, produce an interference indicating the phase of the surface-generated second harmonic.

interfere if they are of opposite phase and constructively interfere if they are the same phase. This beating-against-a-wave-of-known-phase enables determination of the unknown phase.

The first absolute orientation measurement[24] used this interference technique to determine the orientation of phenol on liquid water using SHG as follows. There is a residual fundamental beam reflected along with the SHG signal (Fig. 4). Sending these two beams into a $z$-cut quartz crystal generates a second SHG signal from interaction of the residual fundamental with the quartz crystal. The second signal interferes with the reflected SHG propagating through the crystal. The results indicated that the –OH group of phenol points into the bulk solution.

The principle for absolute orientation with SFG is similar, but the experimental arrangement is more involved. The first SFG phase measurements[22,23] examined pentadecanoic acid, PDA, on water and tetramethyl silane on silica using two configurations. In the first configuration, the residual reflected infrared and visible beams, along with the SF output were refocused onto a $Y$-cut quartz plate where an additional SF beam is generated from the quartz pate. The original and quartz-plate SF beams interfere. In the second experimental arrangement, the adsorbate was deposited directly on the quartz plate. In this arrangement, a SF signal is generated by the PDA along with the nonresonant signal from the quartz. If the system of interest has a significant nonresonant background signal of known phase, this interfere can be exploited to determine the orientation.[30,31]

A recent application of this nonresonant substrate interference technique was reported[31] for water in contact with $\alpha$-quartz (0001). $\alpha$-quartz is birefringent with the optic axis in the surface plane. Rotating the quartz crystal so that the optic axis rotates through the plane of incidence changes the sign of the nonresonant background providing insight into the phase of the broad, hydrogen-bonded region of the water spectrum that extends about 700 $cm^{-1}$ from approximately 2950 $cm^{-1}$. Understanding the origin of the SFG intensity throughout this broad region and changes in various aqueous solutions remains one of the most important challenges for SFG spectroscopy.

### 3.2. *Orthogonal resonances*

The molecular tilt angle reveals information about the interaction of the adsorbate with the surface or adsorbate–adsorbate packing. Hence the focus of numerous SFG studies is determining the tilt angle. Unfortunately, in many cases, the observed intensity is not very sensitive to the tilt angle. Ammonia serves as an example. As shown in Fig. 5, only the *ssp* polarization combination produces significant intensity up to a tilt angle of about 40°. This is nearly half the chemically reasonable angles. Since all intensities, including *ssp* are sensitive to both the orientation angle and the number density, this insensitivity makes it challenging to use SFG to monitor surface concentration for varying conditions.

The broad distribution consistent with one resonance can be narrowed significantly if the molecule possesses a second, nonparallel resonance.[25,26] The first reported example using a nonparallel resonance is that of ammonia on water.[25] The ammonia

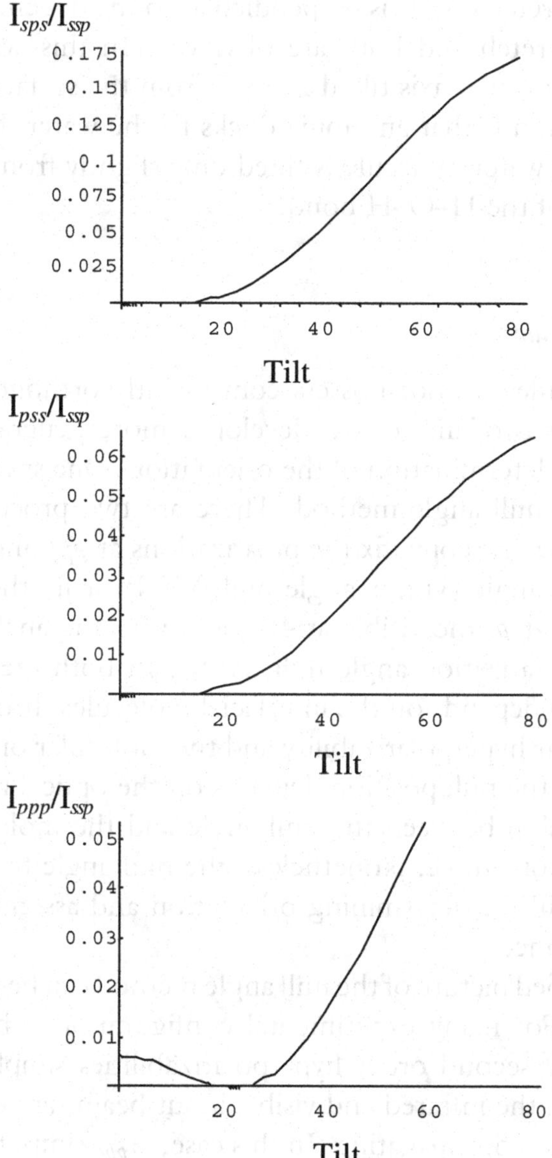


**Fig. 5.** Ammonia symmetric stretch *ppp* intensities relative to that for *ssp* polarization as a function of the tilt of the molecular axis from the surface normal.

symmetric stretch dipole is perpendicular to the degenerate antisymmetric stretch and both are observed. In this system, the ammonia molecular axis tilted 25–38° from the surface normal. The conclusion is that ammonia docks to the water dangling –OH and the water molecule is tilted only slightly from a simple bifurcation of the H–O–H bond.

### 3.3. *Null angle*

Most molecules do not possess conveniently orthogonal resonances, so it is of interest to develop a more general method for accurate determination of the orientation. One such method is called the null angle method. There are two procedures for null angle spectroscopy: fix the polarizations at *ppp* and vary the visible input angle (visible angle null, VAN) or fix the infrared polarization at *p*, the visible at 45°, and vary the analyzer (SF) polarizer (polarization angle null, PAN). In both methods the null position depends on the interfacial molecules through both the molecular hyperpolarizability and the molecular orientation. In addition, the null position depends on the optical factors. So the relationship between the null angle and the molecular orientation is not simple. Nonetheless, the null angle technique is very powerful for determining orientation and assigning vibrational resonances.

A simplified picture of the null angle method can be generated as follows. For many experimental configurations, the expressions for the second order hyperpolarizabilities simplify somewhat. Often, the infrared and visible input beams are coaxial, or nearly so, and copropagating. In this case, $A_{ppp}$ simplifies due to near cancellation of the middle terms in Eq. (8) as follows. From Eq. (8), the sum of the middle terms is

$$[\cos\eta_{r,SF} L_X K_Z^{vis} + \sin\eta_{r,SF} L_Z K_X^{vis}]\chi_{ZXX} K_X^{IR} e_{vis}^{p} e_{IR}^{p} \qquad (16)$$

since $\chi_{ZXX} = \chi_{XZX}$ due to the nonresonant Raman. In the absence of a visible resonance, the index of refraction for the visible and sum frequency beams are equal so the Fresnel and angle factor term, aside from common constants, is

$$\begin{aligned}
&\cos\eta_{\mathrm{r,SF}} L_X K_Z^{\mathrm{vis}} + \sin\eta_{\mathrm{r,SF}} L_Z K_X^{\mathrm{vis}} \\
&\quad = \frac{4n_1^2 n_2 \cos\eta_{i,\omega_1}\cos\eta_{\mathrm{r,SF}}}{(n_{1,\mathrm{SF}}\cos\eta_{t,\mathrm{SF}} + n_{2,\mathrm{SF}}\cos\eta_{\mathrm{r,SF}})} \\
&\qquad \times \frac{[n_1\sin\eta_{\mathrm{r,SF}}\cos\eta_{\mathrm{t},\omega_1} - n_2\cos\eta_{\mathrm{t,SF}}\sin\eta_{\mathrm{t},\omega_1}]}{(n_{1,\omega_1}\cos\eta_{\mathrm{t},\omega_1} + n_{2,\omega_1}\cos\eta_{\mathrm{i},\omega_1})}. \qquad (17)
\end{aligned}$$

Using Schnell's law, the term in square brackets is equal to $n_2[\sin\eta_{\mathrm{t,SF}}\cos\eta_{\mathrm{t},\omega 1} - \cos\eta_{\mathrm{t,SF}}\sin\eta_{\mathrm{t},\omega 1}] = \sin(\eta_{\mathrm{t,SF}} - \eta_{\mathrm{t},\omega 1})$. For a near coaxial, copropagating geometry, the difference in the visible and SF angles is nearly zero since the visible photon carries most of the momentum. The sum in Eq. (17) is thus negligible (and can be minimized by minimizing the angle between the infrared and visible inputs). This near cancellation is particularly helpful since the $K_X^{\mathrm{IR}}$ factor depends on the infrared refractive index, and that changes rapidly near vibrational resonances. Note that for counterpropagating geometry, the angle between the visible and SF beams is large and the term in Eq. (17) cannot be ignored. Running both co- and counterpropagating geometries thus provides a method for determining the index of refraction in the nonlinear layer.[7,20,32]

The simplification of $A_{ppp}$ greatly aids analysis of the intensity as a function of either the visible input angle, $\eta_{\mathrm{vis}}$, or the sum frequency polarization direction, $\Theta$ (measured from the output plane). With the simplification $A_{ppp}$ becomes

$$A_{ppp} = [\cos\eta_{\mathrm{r,SF}} L_X \chi_{XXZ} K_X^{\mathrm{vis}} + \sin\eta_{\mathrm{r,SF}} L_Z \chi_{ZZZ} K_Z^{\mathrm{vis}}] K_Z^{\mathrm{IR}} e_{\mathrm{vis}}^p e_{\mathrm{IR}}^p. \qquad (18)$$

Note that the product of Fresnel factors, $L_X K_X^{\text{vis}}$, is the opposite sign from the product $L_Z K_Z^{\text{vis}}$ due to phase reversal of the sum frequency $X$ component relative to the visible input at the interface. These opposite signs result in an interference between $\chi_{XXZ}$ and $\chi_{ZZZ}$ that often leads to a lower than expected *ppp* intensity.

$$\begin{aligned}\frac{A_{ppp}}{K_Z^{\text{IR}} e_{\text{vis}}^{p} e_{\text{IR}}^{p}} &= \cos\eta_{\text{r,SF}} L_X \chi_{XXZ} K_X^{\text{vis}} + \sin\eta_{\text{r,SF}} L_Z \chi_{ZZZ} K_Z^{\text{vis}} \\ &= \frac{-4n_1^2 \cos\eta_{\text{i},\omega 1} \cos\eta_{\text{r,SF}}}{(\cos\eta_{\text{t,SF}} + n_{\text{SF}} \cos\eta_{\text{r,SF}})(\cos\eta_{\text{t},\omega 1} + n_1 \cos\eta_{\text{i},\omega 1})} \\ &\quad \times \left[\cos\eta_{\text{t,SF}} \cos\eta_{\text{t},\omega 1} \chi_{XXZ} - \frac{n_1}{(n')^4} n_2^3 \sin\eta_{\text{r,SF}} \sin\eta_{\text{t},\omega 1} \chi_{ZZZ}\right].\end{aligned} \tag{19}$$

The interference is exploited in null angle methods. It can be accessed either by changing the visible input angle (VAN, Sec. 3.3.1) or by rotating the sum frequency analyzer polarizer (PAN, Sec. 3.3.2).

### 3.3.1. *Visible angle null, VAN*

Equation (19) indicates that the balance between $\chi_{XXZ}$ and $\chi_{ZZZ}$ depends on the transmitted and reflected sum frequency angles, $\eta_{\text{t,SF}}$ and $\eta_{\text{r,SF}}$ respectively, and the transmitted visible angle, $\eta_{\text{t},\omega 1}$. Figure 6 shows a plot of $A_{ppp}$ as a function of the visible input angle (omitting the common $K_Z^{\text{IR}}$ factor and normalizing with respect to the input visible and infrared fields). The interference between $\chi_{XXZ}$ and $\chi_{ZZZ}$ is a function of their magnitudes. The magnitudes, in turn, are a function of the interfacial molecules and their orientation. The case where $2|\chi_{XXZ}| = |\chi_{ZZZ}|$ is plotted in Fig. 6. For this case, the *ppp* intensity vanishes for a visible input angle of 38.1°: the visible

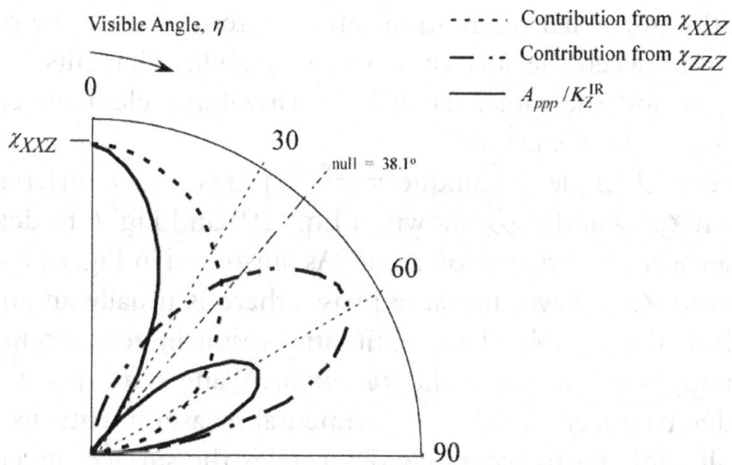


**Fig. 6.** Illustration of the effect of the visible input angle on the balance between $\chi_{XXZ}$ and $\chi_{ZZZ}$ in the *ppp* polarization combination for the case where $|\chi_{ZZZ}| = 2|\chi_{XXZ}|$. Due to the negative sign in Eq. (19), the contributions ***destructively interfere*** when they are of the same phase and ***add*** (not shown) when the $\chi_{XXZ}$ and $\chi_{ZZZ}$ susceptibilities are the opposite phase. This simulation uses unit infrared and visible intensities, a substrate index of refraction of 1.33, and an infrared input angle of 60°.

angle null, VAN = 38.1°. Note that Eq. (19) can be used to determine the optimum visible input angle for each system.

Using Schnell's law to bring all angles into the second medium, Eq. (19) indicates that the null angle occurs when

$$\cos\eta_{t,SF}\cos\eta_{t,\omega_1}\chi_{XXZ} = \frac{n_2^4}{(n')^4}\sin\eta_{t,SF}\sin\eta_{t,\omega_1}\chi_{ZZZ} \tag{20}$$

or

$$\frac{n_2^4}{(n')^4}\tan\eta_{t,SF}\tan\eta_{t,\omega_1} = \chi_{XXZ}/\chi_{ZZZ}. \tag{21}$$

The sum frequency angle is often near to the visible angle, so the null angle is approximately

$$\eta_{t,null} = \frac{(n')^2}{n_2^2}\arctan\sqrt{\chi_{XXZ}/\chi_{ZZZ}}. \tag{22}$$

Thus, the molecular orientation can be determined via the relationship between the surface hyperpolarizability elements $\chi_{XXZ}$ and $\chi_{ZZZ}$ and the molecular hyperpolarizability elements contributing to the resonance.

The null angle technique[7,27,28] exploits the interference between $\chi_{XXZ}$ and $\chi_{ZZZ}$ shown in Eq. (19) and Fig. 6 to determine an accurate orientation angle. As illustrated in Fig. 6, when $\chi_{XXZ}$ and $\chi_{ZZZ}$ have the same phase, there is usually an angle for which the combined susceptibilities result in zero intensity. This angle is referred to as the *null angle*. Wang *et al.*[7] point out that, due to uncertainties in experimental measurements, use of the null angle more accurately determines the surface susceptibilities and thus results in less uncertainty in determination of the molecular orientation.

On most surfaces, thermal energy combines with adsorbate binding strength to result in a distribution of molecular orientations except at the very lowest temperatures or for very directed bonding between the adsorbate and the surface. Thus, determining the orientation distribution reveals data about the adsorbate–surface interaction. Since the null angle technique determines the average orientation, it is of interest to investigate the effect of an orientation distribution on the null angle data. An orientation distribution affects the null angle via the relationship between the molecular hyperpolarizability, $\beta^{(2)}$, and the surface hyperpolarizability, $\chi^{(2)}$, so the relationship is somewhat convoluted. The example chosen to illustrate the effect of an orientation distribution is the following. The molecule is pyramidal with a polarizability along the molecular axis, $\beta_{ccc}$, that is 3.5 times larger than that perpendicular to the axis, $\beta_{aac} = \beta_{bbc}$, and all off diagonal polarizabilities are much smaller. The comparison shown in Fig. 7 is between such a molecule oriented with a delta-function distribution at 35° and a bimodal distribution with the same average orientation, specifically at 20° and 50°. The delta-function distribution results in a null angle of 39.1°

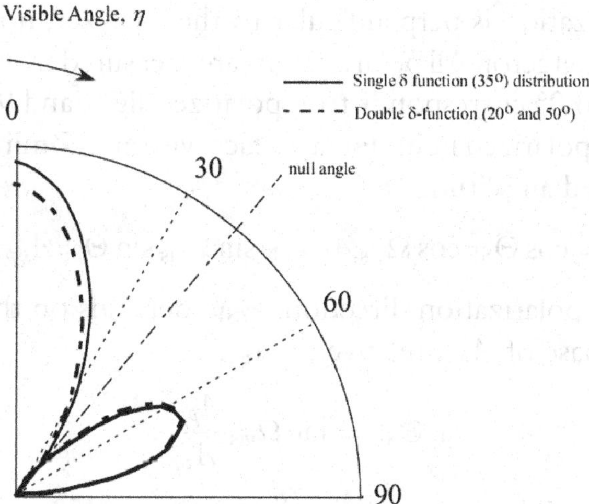


**Fig. 7.** A distribution of molecular angles alters the relationship between $\chi_{XXX}$, and $\chi_{ZZZ}$. This example compares $A_{ppp}$ for two different delta function distributions: a single $\delta$-function at 35° and two delta functions at 20° and 50°; $|\beta_{ccc}| = 3.5\ |\beta_{aac}|$, and molecular hyperpolarizability related to the surface hyperpolarizabilty as given in Eq. (10). Substrate index of refraction = 1.33, $\eta_{i,IR} = 60°$.

and the bimodal distribution has a null angle of 38.3°. The subtle difference results from the gentle variation of $\chi_{XXZ}$ and $\chi_{ZZZ}$ with molecular orientation. The null angle becomes increasingly sensitive to the distribution as the molecular axis polarizability increases relative to that of the perpendicular axis. Thus, the molecular system determines the sensitivity of the null angle to an orientation distribution.

### 3.3.2. *Polarization angle null, PAN*

For some experimental constraints it is not possible to alter the visible input angle significantly. Fortunately, there is an equivalent method to determine $\chi_{XXZ}$, $\chi_{ZZZ}$ and their relative phase: choose an input angle that generates significant *ppp* intensity, fix the IR polarization at *p*, the visible at 45°, and vary the output polarization analyzer. When the generated sum frequency beam polarization is perpendicular to the analyzer, no intensity reaches the detector. All polarizations are measured from the output plane, so 0° corresponds to $p$-polarized light and 90° corresponds to $s$-polarized light. For a surface with an $XZ$ mirror plane, the generated amplitude is

$$A_{\Theta\Omega p} = \cos\Theta_{SF}\cos\Omega_{vis}A_{ppp} + \sin\Omega_{vis}\sin\Theta_{SF}A_{ssp}. \tag{23}$$

The output polarization direction, $\Theta_{SF}$, depends on the magnitude and phase of $A_{ppp}$ relative to $A_{ssp}$.

$$\tan\Theta_{SF} = \tan\Omega_{vis}\frac{A_{ppp}}{A_{ssp}}. \tag{24}$$

Setting the visible input polarization at $\Omega_{vis} = 45°$ generates an output polarization angle equal to $\arctan(A_{ppp}/A_{ssp})$. The null angle is perpendicular to this direction.

The dependence on the surface polarizability, $\chi_{XXZ}$ and $\chi_{ZZZ}$, is thus

$$\begin{aligned} A_{\Theta 45p} = \{&[\cos\Theta\cos\eta_{r,SF}L_X K_X^{vis} + \sin\Theta L_Y K_Y^{vis}]\chi_{XXZ} \\ &+ \cos\Theta\sin\eta_{r,SF}L_Z K_Z^{vis}\chi_{ZZZ}\}\cos(45°)K_Z^{IR}\, e_{vis}^{45}\, e_{IR}^{p}. \end{aligned} \tag{25}$$

For $\Theta = \pm 90°$, the signal depends only on $\chi_{XXZ}$. The angle at which the $\chi_{XXZ}$ contribution vanishes depends on the Fresnel factors:

$$\tan\Theta = -\frac{\cos\eta_{r,SF}L_X K_X^{vis}}{L_Y K_Y^{vis}}, \tag{26}$$

and thus is a property of the substrate and the input angle. For a visible angle of 50°, the $\chi_{XXZ}$ contribution vanishes at $\Theta = 35.4°$.

Since $L_X$ is negative, the coefficient of $\chi_{XXZ}$ changes sign between $\Theta = 0°$ and $\Theta = 90°$ starting negative, going to zero at the angle $\Theta$ given in Eq. (26), and becoming positive at larger $\Theta$. Since the coefficient of $\chi_{ZZZ}$ is positive and diminishes as $\cos\Theta$, whether there is a polarization direction in the first quadrant for which the intensity vanishes depends on the magnitude of $\cos\eta_{r,SF}L_X K_X^{vis}\chi_{XXZ}$ compared with $L_Z K_Z^{vis}\chi_{ZZZ}$. If the former is larger than the latter, there is a null angle in the first quadrant. The magnitude of both factors depends on the optical constants of the substrate through the Fresnel factors and on the orientation of the molecule(s) in the interfacial region via $\chi_{XXZ}$ and $\chi_{ZZZ}$. An example is shown in Fig. 8 for which PAN is equal to $-27.5°$ for a molecule with orientation such that $2|\chi_{XXZ}| = |\chi_{ZZZ}|$.

It is important to carefully define the coordinate system to fully analyze the polarization data. The $X$ axis is in the surface and the positive direction is taken as the forward direction for the visible and infrared beams. The $Z$ axis is perpendicular to the surface, so the positive $Y$ axis is to the left when facing in the forward direction of the infrared and visible beams. The zero angle for the polarization is taken as the $Z$ axis: that is $p$-polarized incoming light. With these axes, the first quadrant is to the left and the fourth quadrant is to the right when facing in the forward

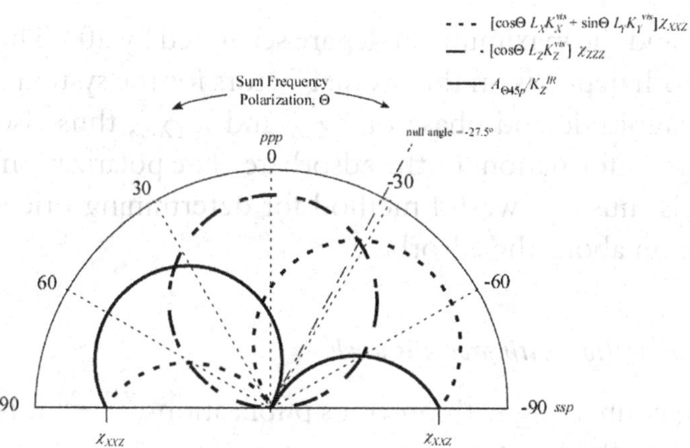


**Fig. 8.** On a dielectric surface with $n = 1.33$, the contribution of the $\chi_{XXZ}$ to the intensity vanishes when $\Theta = 35.4°$ due to the opposite phases of the $L_X K_X$ and $L_Z K_Z$ Fresnel factors. The null angle is $-27.5°$. Simulation parameters: $\chi_{ZZZ} = 2\chi_{XXZ}$, $\eta_{i,vis} = 50°$.

direction of the beams. Rotating to the left — counterclockwise — is a +45° rotation; rotating to the right — clockwise — is a −45° rotation. The plot in Fig. 8 is based on a +45° rotation, with a −45° rotation, the plot is flipped left to right.

It is helpful to rearrange Eq. (25) as

$$\frac{A_{\Theta 45p}}{K_Z^{\mathrm{IR}} e_{\mathrm{vis}}^{45} e_{\mathrm{IR}}^{p} \cos(45°)} = \cos\Theta[\sin\eta_{\mathrm{r,SF}} L_Z K_Z^{\mathrm{vis}} \chi_{ZZZ} + \cos\eta_{\mathrm{r,SF}} L_X K_X^{\mathrm{vis}} \chi_{XXZ}] + \sin\Theta L_Y K_Y^{\mathrm{vis}} \chi_{XXZ}. \tag{27}$$

Then the null angle is

$$\tan\Theta_{\mathrm{null}} = -\frac{\sin\eta_{\mathrm{r,SF}} L_Z K_Z^{\mathrm{vis}} \chi_{ZZZ} + \cos\eta_{\mathrm{r,SF}} L_X K_X^{\mathrm{vis}} \chi_{XXZ}}{L_Y K_Y^{\mathrm{vis}} \chi_{XXZ}}. \tag{28}$$

and the maximum intensity is at a polarization angle of

$$\cot\Theta_{\max} = \frac{\sin\eta_{\mathrm{r,SF}} L_Z K_Z^{\mathrm{vis}} \chi_{ZZZ} + \cos\eta_{\mathrm{r,SF}} L_X K_X^{\mathrm{vis}} \chi_{XXZ}}{L_Y K_Y^{\mathrm{vis}} \chi_{XXZ}} \tag{29}$$

The null and the maximum angles are separated by 90°. The position of both depends on the Fresnel factors for the system as well as the magnitude and phase of $\chi_{ZZZ}$ and $\chi_{XXZ}$, thus also contains phase information for the adsorbate. The polarization angle method is thus a powerful method for determining orientation information about the adsorbate.

### 3.3.3. *Connection with previous work*

To aid in connecting with previous publications,[7,27,28] it is helpful to define the functions

$$A(\eta) = \frac{\cos\eta_{\mathrm{r,SF}} \cos\eta_{\mathrm{i},\omega_1} \cos\eta_{\mathrm{t,SF}} \sin\eta_{\mathrm{t},\omega_1}}{(\cos\eta_{\mathrm{t,SF}} + n_{2,\mathrm{SF}} \cos\eta_{\mathrm{r,SF}})(\cos\eta_{\mathrm{t},\omega_1} + n_{2,\mathrm{vis}} \cos\eta_{\mathrm{i},\omega_1})} \tag{30}$$

and

$$B(\eta) = \frac{n_2 \sin \eta_{r,SF} \cos \eta_{i,\omega_1}}{(n')^2(\cos \eta_{t,SF} + n_{2,SF} \cos \eta_{r,SF})(\cos \eta_{t,\omega_1} + n_{2,vis} \cos \eta_{i,\omega_1})} \times \cos \eta_{r,SF} \sin \eta_{t,\omega_1}. \quad (31)$$

With the above-listed definitions, $A_{ppp}$ for the symmetric stretch of ammonia becomes

$$\begin{aligned} \frac{A_{ppp}}{K_Z e^p_{vis} e^p_{IR} 4n_2^2} &= A(\eta)\chi_{XXZ} + B(\eta)\chi_{ZZZ} \\ &= [(1/2)A(\eta)(\beta_{aac} + \beta_{ccc}) + B(\eta)\beta_{aac}] \cos\theta \\ &\quad + (1/2)A(\eta) - B(\eta)](\beta_{aac} - \beta_{ccc})\} \cos^3\theta. \end{aligned} \quad (32)$$

The parameter $c_0$ defined in the publications of Wang *et al.*[7,27,28] is then

$$c_0 = \frac{[(1/2)A(\eta) - B(\eta)](\beta_{ccc} - \beta_{aac})}{(1/2)A(\eta)(\beta_{aac} + \beta_{ccc}) + B(\eta)\beta_{aac}} \quad (33)$$

and the functional $r(\theta) = \cos\theta - c_0 \cos^3\theta$ is related to $A_{ppp}$ via

$$\frac{A_{ppp}}{K_Z e^p_{vis} e^p_{IR} 4n_2^2} = [(1/2)A(\eta)(\beta_{aac}+\beta_{ccc})+B(\eta)\beta_{aac}]r(\theta). \quad (34)$$

As pointed out by Wang *et al.*, $I_{ppp} = 0$ when $\cos\theta = c \cos^3\theta$ or $c_0 = 1/[cos^2\theta]$. Thus, finding the null angle, VAN, determines the molecular tilt angle. Further, given experimental uncertainties in determining intensities, the tilt angle can be more accurately determined with the null angle method.

### 3.3.4. *Example*

The preceding discussion is very general. To illustrate the methods, it is useful to apply them to a specific molecular system.

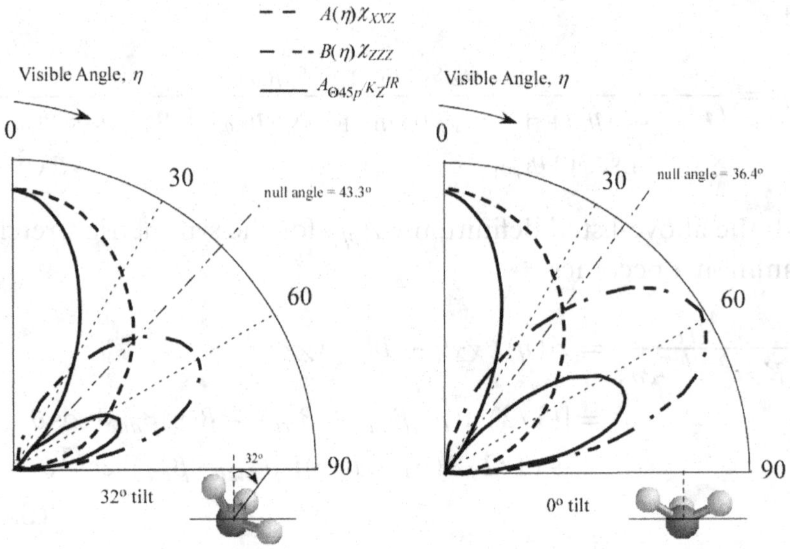


**Fig. 9.** $A_{ppp}$ as a function of visible input angle for ammonia on a dielectric surface, $n = 1.33$, infrared angle = 46°, visible polarization fixed at *ppp*. (a) δ-function distribution at 32°, (b) δ-function distribution at 0°. In (a), the null angle is 43.3°, while in (b) the null is at 36.4°.

In this section, VAN and PAN are illustrated for the symmetric stretch of ammonia on a dielectric surface with $n = 1.33$, $\eta_{i,IR} = 46°$, and for PAN, $\eta_{i,vis} = 52°$.

For the symmetric stretch of ammonia, the variation of $A_{ppp}$ with input angle is illustrated in Fig. 9. With the ammonia $C_3$ axis tilted 32° from the normal, VAN is 43.3°. With ammonia sitting vertically, VAN is 36.4°. For both configurations of ammonia, the largest signal is obtained with a very steep input angle.

The polarization rotation method can also be illustrated with ammonia. Figure 10 illustrates the result of rotating the analysis polarizer with the visible input angle fixed at 52° and polarization at 45°; the infrared input angle at 46° and *p* polarized. Figure 10 illustrates the importance of the molecular hyperpolarizabilities. Compare Fig. 10 with the similar plot in Fig. 8: PAN −35.4°

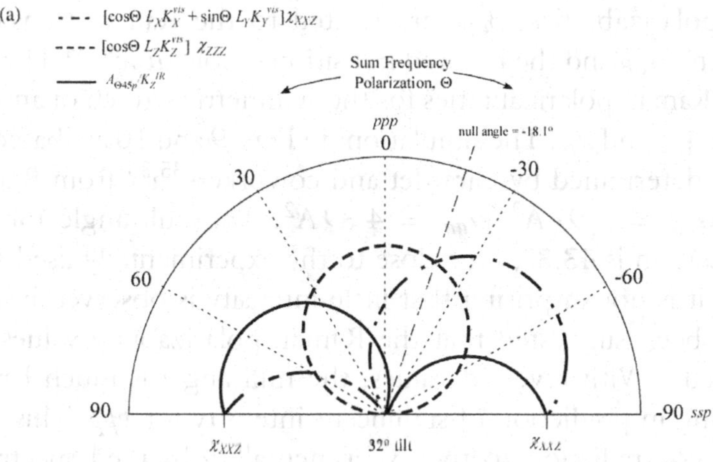


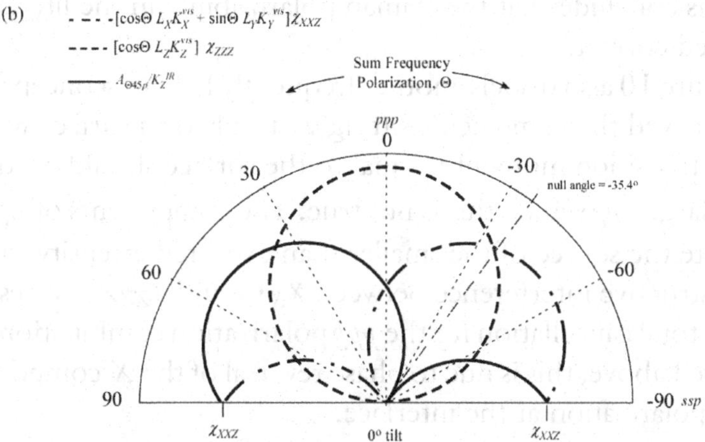


**Fig. 10.** Polarization angle dependence of $A_{ppp}$ for ammonia on a dielectric surface with $n$ = 1.33, fixed visible input angle = 52°, and fixed infrared angle = 46°. (a) $\delta$-function distribution of the $C_3$ axis at 32°, (b) $\delta$-function distribution of the $C_3$ axis at 0°. Notice that the contribution of the $\chi_{XXZ}$ term vanishes at a polarization angle of 35.1°.

(near to *ppp*) when ammonia sits vertically on the surface and −18.1° when tilted at 32°, while the parameters used in Fig. 8. produce a clear PAN = −27.5°.

The simulation in Fig. 9 also clears up a controversy in the literature[7] as follows. For vibrational SFG, the molecular hyperpolarizabilities, $\beta_{\alpha\beta\gamma}$ are related to the Raman transition moment, $\alpha_{\alpha\beta}$ and the infrared transition dipole, $\mu_{\gamma}$.[33,34] The relevant Raman polarizabilities for the symmetric stretch of ammonia are $\alpha_{aa}$ and $\alpha_{cc}$. The simulations in Figs. 9 and 10 are based on values determined by Girardet and coworkers[35,36] from Raman data, $\alpha_{aa} = 1.97\,\text{Å}^2$, $\alpha_{aa} = 4.39\,\text{Å}^2$. The null angle for *ppp* polarization is 43.3°, very close to the experimentally used 46°. Thus, it is not surprising that little intensity is observed in *ppp*. It has been suggested that the Raman polarizability values are reversed.[7] With reversed values, the null angle is much larger, resulting in prediction of significant intensity for *ppp*. This is in direct contradiction to the experimentally observed spectra.[25] We thus conclude that the Raman polarizability in the literature is indeed correct.

Figure 10 also dispels another frequently held misconception. It is believed that a molecule sitting vertically on a surface with a strong transition moment normal to the surface should produce a very large *ppp* signal: this is not true. The components of $\chi^{(2)}_{\Theta 45p}$ illustrate the source of the smaller than expected intensity: there is a destructive interference between $\chi_{XXZ}$ and $\chi_{ZZZ}$ that results in near total cancellation for the *ppp* polarization combination. As indicated above, this is due to phase reversal of the $X$ component of the polarization at the interface.

## 4. Current Issues in Sum Frequency Generation

### 4.1. *Interfacial optical constants and bulk contributions*

In recent years, SFG has been used to probe more and more complex systems. As the technique gets applied to these complex environments, it becomes increasingly important to experimentally verify — or challenge — some of the assumptions built into the above-discussed foundations. One of the assumptions concerns the Fresnel and optical factors for the interfacial layer, $n'$

in Table 3. In most cases the optical constants for the interfacial layer are not known or are ill defined, particularly in the limit of a molecularly thin monolayer. These constants are not just of academic interest: different assumptions for the optical constants can result in deduced orientations for the interfacial molecules that differ by several degrees.[37] A widely used approximation consists of assigning the interfacial layer a refractive index that is equal to one of the two bounding media.[38–42] In other cases, the interfacial layer is assigned an index that is the average of the two bounding media.[20,37,43–46] In still other cases, the index is estimated from ellipsometry measurements.[47,48]

Recently, Simpson *et al.*[32] have developed a theoretical basis for the optical constants of the interfacial layer in which propagation of the nonlinear beam within the layer is described as in waveguide propagation. The polarization leaks across the edges of the waveguide; the amplitude and slope of the wave are continuous across the boundary. In the limit of a thin layer, the waveguide becomes lossy due to emission of the reflected and transmitted sum frequency beam. Within this model, the refractive index at any point contains contributions from both media, each weighted by an exponential decay. The refractive index then depends on the thickness, $d$, of the film relative to the decay length, $1/\gamma$. In the limit, where the interfacial layer thickness is much smaller than the wavelength of the incident light: $1/\gamma \ll \lambda$, i.e., the ultra-thin-film limit, the refractive index of the film is equal to the average of the indices of the two bounding media. The thin-film limit applies not only to a molecular monolayer, but also to any surface where the surface structure differs from that of the bulk.

To test the waveguide treatment, Simpson *et al.* developed a nonlinear optical null ellipsometry (NONE) technique[49] to measure the nonlinear tensor elements. The instrument uses two detectors and several waveplates to determine the simultaneous null angles for right circular, left circular, $+45°$, and $-45°$

incident light. Combinations of these angles determine the relative values of the $\chi_{xxz}$, $\chi_{zxx}$, and $\chi_{zzz}$ tensor elements plus their phases. In the systems tested to date, the general result is that the index of refraction of the interfacial layer is the average of the two bounding media. This is the approximation used to generate Figs. 6–10. Further work is needed to determine the generality of this result.

Related to the issue of the nonlinear optical constants in the surface is the issue of surface versus bulk contributions to the signal. Within the electric dipole approximation, a centrosymmetric media does not contribute to the SHG or SFG signal. The higher order terms, the bulk quadrupole and magnetic dipole, do contribute to the signal and since they accumulate over a larger number of scatterers, the higher order contributions to the signal can be as large as the surface electric dipole contribution. Shen and coworkers[50] have examined this issue by comparing SFG in reflection to that in transmission. Since the coherence length in transmission is quite long, the bulk contribution is strongly enhanced in transmission compared with that in reflection. For a monolayer of octadecyltrichlorosilane (OTS) on fused quartz, the reflected and transmitted SFG signals are well correlated after the nonresonant signal is accounted for. (The nonresonant part is deduced from the opposite phases of the Fresnel factors in the two cases.) It should be noted that in this experimental arrangement, OTS is not expected in the bulk, hence any bulk contribution originates from the fused quartz substrate.

In contrast to OTS on quartz, many SFG studies are done on systems where the molecule responsible for the signal is also located in the bulk. Lack of a bulk contribution then depends on the bulk having isotropic symmetry and the validity of the electric dipole approximation. Examination of a polyethylene film in both reflection and transmission[50] shows that there is indeed a quadrupolar contribution to the signal. Specifically, the methylene antisymmetric stretch, which is Raman active but IR inactive appears in the transmission spectrum but not in the reflection spectrum. This mode gains oscillator strength in transmission due to IR quadrupolar excitation.

Comparison of reflection versus transmission signals is a powerful method for determining the bulk quadrupolar contribution. The major limitation on this method is the requirement that both the infrared and visible beams must be able to penetrate into the bulk of the sample under investigation. To date, this limitation has not been overcome for aqueous samples for which the strong infrared absorption quenches generation of a bulk signal.

### 4.2. *Collective modes — a theoretical challenge*

Since water and aqueous solutions are both ubiquitous and extremely important, there have been numerous studies of aqueous interfaces. One of the major challenges in analyzing the SFG spectra from such interfaces is interpreting the spectrum of water (Fig. 11), which extends for more than $700\,\text{cm}^{-1}$. The excessive width is due to coupling between water molecules. To appreciate the importance of coupling, considering the following. In the gas phase intramolecular coupling of the OH

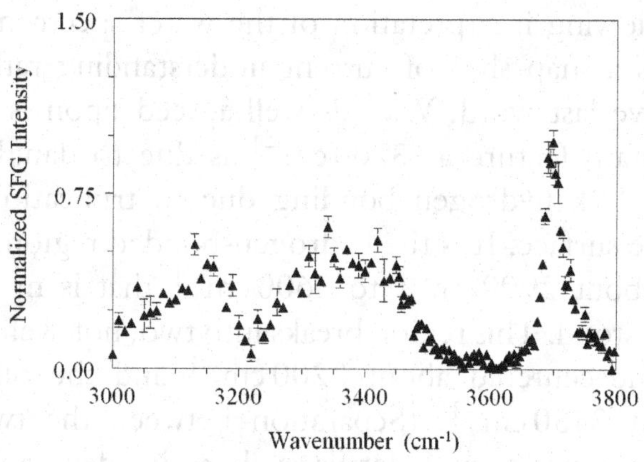


**Fig. 11.** The *ssp* spectrum of water with incident angles of 46° (IR) and 52° (vis).

oscillators gives rise to the well-known symmetric and asymmetric stretches that are separated by about 100 $cm^{-1}$. In the condensed phase environment the oxygen electrons that provide the gas-phase coupling are involved in hydrogen bonding, reducing the importance of intramolecular coupling.[17,51] Water has a significant transition dipole, so neighboring water molecules couple and this intermolecular coupling is large[18,51] leading to in-phase and out-of-phase oscillations that spread the resonances over hundreds of wavenumbers and boost the intensity by orders of magnitude.[17,52] Nonetheless, most attempts to reproduce the SFG spectra of aqueous interfaces have neglected intermolecular coupling,[53–56] and many interpretations of water spectra discuss resonances in terms of independent molecules.[57–59]

Treating significant intermolecular coupling in a liquid system is a particularly challenging theoretical endeavor. The current state of the art treats the coupled OH dipoles as coupled Morse oscillators with a polarizable intermolecular potential.[17,18,51] The local field parallel to the OH bond is averaged over a classical trajectory for several *p* sec. with periodic boundary conditions. The computed spectra do not quantitatively agree with the experimental ones, but they do reproduce the major features and provide insight into the dynamics at the aqueous interface.[18]

The following interpretation of the water spectrum should be taken as a snap shot of current understanding rather than the definitive last word. What is well-agreed upon is that the relatively sharp feature at 3700 $cm^{-1}$ is due to dangling OH bonds that lack hydrogen bonding due to truncation of the liquid at the surface. It is the hydrogen-bonded region, extending from about 2900 $cm^{-1}$ to 3600 $cm^{-1}$ that is more tenuously understood. This region breaks into two, not-well resolved features: one centered about 3200 $cm^{-1}$ and the other centered about 3450 $cm^{-1}$. (Separation between the two peaks depends on experimental details such as incident angles and temperature.)

One strategy for unraveling the origin of the spectral features consists of adding solutes and determining how they perturb the neat water spectrum. At this point, a number of soluble salt solutions have been examined with the general result that large, singly charged anions result in an enhanced intensity in the 3450 $cm^{-1}$ peak.[58] Acid solutions result in a much more significant enhancement of the 3200 $cm^{-1}$ band.[60–64]

Treating the solution as a large system of coupled oscillators provides the following insight into the origin of these two broad peaks and perturbation of the peaks with addition of solutes. The less red-shifted band at 3450 $cm^{-1}$ appears to have a major contribution from the top bilayer. Hydrogen bonds in the top bilayer are strained and thus relatively weak. Hence these are found within the less red-shifted band. These hydrogen bonds include three coordinate water with a dangling lone pair as well as four-coordinate water in the lower half bilayer. On average, the O–H dipole of these water molecules points out of the solution.[18] Large anions penetrate to this top bilayer[58] tightening the orientation distribution and increasing the intensity. Three coordinate water molecules with a dangling H, i.e., free–OH waters, appear to have very weak resonances and contribute little to the SFG intensity.

The more red-shifted band at 3200 $cm^{-1}$ has contributions from several bilayers, extending far into the solution (estimated to be 4–7 layers). Collective modes dominate with the most symmetrical combinations being the most red shifted; the bluer end of the band being due to less symmetrical combinations. The net O–H dipole points out of the solution due to a tighter orientation distribution of those OH bonds that point into the solution relative to those that point out. Acid solutions modify this band as follows. Computational studies indicate that $H_3O^+$ has a propensity for the top monolayer of aqueous solutions[58,65,66] due to the oxygen of $H_3O^+$ being a weak H-bond acceptor. (Section 5.1. below presents a broader discussion of this issue.) The $H_3O^+$

dipole points out of the solution and, due to Coulombic interaction, tightens the orientation distribution of the subsurface dipoles. An increased dipole projection onto the surface normal increases the SFG oscillator strength. Note that $H_3O^+$ does not directly contribute to this band: it is very broad.[67]

Based on insights gained from including intermolecular coupling in the theoretical models, it is expected that further progress depends on increasingly sophisticated treatments of collective modes in larger and larger systems. It is expected that the above discussion of the model for water and aqueous solutions will evolve due to increasingly sophisticated theories and experiments.

### 4.3. *Probe depth*

SFG derives its surface specificity from the asymmetry of the interface; however, the question of how deep into the surface that asymmetry persists is still an open question. The origin of the uncertainty can be appreciated with a discussion of length scales. The coherence length of the sum frequency process is on the order of the wavelength of the shorter wavelength excitation; usually hundreds of nanometers. The molecular scale is on the order of a few tenths of a nanometer; the coherence length is orders of magnitude longer than memory of the surface truncation is likely to persist. Thus, SFG gives no direct information on the probe depth. Nonetheless, where a molecule sits within the interface, how it is oriented, and the structure of the solvation shell are all important issues for kinetic processes at the interface as well as for electrochemical processes.[68] Getting this important information from the data requires augmenting the SFG process. Two such augmentations are discussed below: using a solvatochromic chromophore to probe the local polarity[69–71] and using the Stark effect shifting of a vibrational resonance to determine the screening depth of the surface change on an electrode by the solvent.[72]

Imagine a molecule that consists of a hydrophobic, solvent-sensitive chromophore attached to a polar or charged head group by an alkyl chain of specified length. At a hydrophobic–hydrophilic interface, the hydrophobic chromophore floats into the hydrophobic phase while the charged or polar head group remains anchored in the hydrophilic phase. The chromophore thus is located at varying distances from the interface; a distance that depends on the length of the alkyl chain connecting the ends. Solvatochromism occurs because of a difference in polarity of the chromophore in its ground and excited states. If the excited state dipole is larger than that of the ground state, the excited state is stabilized in a polar environment and the excitation spectrum red-shifts with respect to the gas-phase spectrum of the same chromophore. The red shift is increasingly pronounced with increasing solvent polarity. The chromophore thus acts as a reporter of the local polarity. One such chromophore is *p*-nitroanisole, which has a 20 nm red shift in going from cyclohexane to water.[73] The assemblage of the solvatochromatic probe, alkyl spacer, and head-group is dubbed a molecular ruler[70,71] and has a resolution of a methylene spacer.

Using the molecular ruler to probe polarity[70,71] results in the following conclusions for organic alcohols. (a) The interface is molecularly flat and quite sharp. (b) For a butanol–hydrophilic interface, the interfacial polarity is greater than that of bulk butanol. The origin of this greater polarity is not yet explained, but is probably related to the molecular ordering at the interface. (c) Surface polarity effects extend no more than one solvent layer into the solution. To date this work has not been extended to other solvents, so further developments are required prior to drawing general conclusions about polarity changes across interfaces.

The second approach to determine interfacial width[69–72] takes advantage of the vibrational Stark effect. Infrared active vibrational modes necessarily have a change in dipole upon excitation. A dipole is a charge separation, hence this separation can be affected by an applied electric field. For CO on Pt, the local electric Stark shift, $d\nu_{CO}/dE_{local}$, is $1 \times 10^{-6}\,\mathrm{cm}^{-1}/(\mathrm{V/cm})$[74] and the Stark tuning rate, $d\nu_{CO}/d\phi$ is $30\text{–}45\,\mathrm{cm}^{-1}/\mathrm{V}$.[75–78] Combining these data for CO at an electrochemical interface gives[72]

$$d = \frac{\frac{d\nu_{CO}}{dE_{local}}}{\frac{d\nu_{CO}}{d\phi}} = 330\,\mathrm{pm} \tag{35}$$

suggesting that the double layer is one ion layer thick; a result consistent with capacitance data. CO is a particularly good choice for this measurement as the Stark shift is large.

These two methods of measuring the thickness of the probed layer are consistent with the general statement that SFG probes 1–3 monolayers. It should be noted that the probed thickness probably varies with the system and the solutes (for example, the model for water discussed in Sec. 4.2 suggests that the ionic solution interface has contributions at least to 7 monolayers). A more definitive answer concerning the probe depth awaits both theoretical and experimental developments.

### 4.4. *Nanoparticle SFG*

The origin of the surface specificity of SHG and SFG is that the interface is an inherently noncentrosymmetric environment, leading to the assumption that any centrosymmetric environment cannot give a signal. The origin of this notion is that the nonlinear polarization induced in oppositely oriented molecules are of opposite phase and thus cancel resulting in no signal. This is why there is no second harmonic or sum frequency signal from the bulk. However, if the oppositely oriented molecules are separated by a distance long enough that there is a phase shift between the signals generated by the molecules at each of the surfaces, there is a net signal. Thus, coherent second harmonic signals have been observed from several systems including polystyrene microspheres, micelles, and nanoparticle in solution.[6,79]

When the spherical particle becomes even smaller, smaller than the wavelength of the excitation light, then the phase matching becomes somewhat more complex involving Rayleigh–Gans–Debye scattering theory.[80–85] The forward scattered sum frequency or second harmonic signal vanishes due to the opposite phases on each side of the spherical particle. In the nonforward directions, the phase mismatch across the sphere breaks inversion symmetry — the phase mismatch is nonzero — and the scattered intensity is thus also nonzero. This leads to an angular pattern to the nonlinear intensity with a maximum shifting toward 90° from the forward direction as the particles get smaller. The Rayleigh–Gans–Debye theory has been experimentally confirmed for polystyrene beads as small as 55 nm.[81]

### 4.5. *Time resolution*

The vast majority of SFG studies to date have examined static, mostly equilibrium systems. SFG is not inherently limited to static systems, it is also capable of probing ultrafast surface dynamics: all that is required is a sufficiently short pulse. The experimental scheme consists of splitting the laser output into two beams: one to pump an excited state and the second to probe with either SHG or SFG. The first such studies used SHG to probe relaxation in rhodamine 110 adsorbed on silica.[86] Two relaxation times were found, and the photophysics of the adsorbed dye differed greatly from that in solution: a result consistent with formation of dimers at the solid surface. Subsequent studies probed solvent reorganization and rotational dynamics at the aqueous–air interface.[87–90] It was found that the surface reorientation time is slower than that in the bulk and that it is anisotropic. The increased interfacial friction is presumably due to dynamics in the H-bonded network at the surface, which differ from those in the bulk.

In more recent developments, vibrational relaxation at the aqueous interface has been probed using both time and frequency domain spectroscopy.[91,92] Population relaxation in the dangling–OH occurs with a time constant of 1.3 ps, while the H-bonded peaks relax on a faster time scale. Spectral diffusion experiments on the hydrogen-bonded peaks have just begun and likely will form a cornerstone of a fuller understanding of interactions and dynamics at the aqueous surface.

### 4.6. *Surface 2D imaging*

Nonlinear spectroscopy provides excellent resolution perpendicular to the surface normal: on the molecular scale to a depth where the influence of surface truncation dampens. Developments to achieving *xy* resolution within the surface layer are challenged by the weakness of the SF or SH signal — much courser than the diffraction limit of the visible excitation or generated sum frequency. To increase signal intensity, the first reported 2D resolved results enhanced the signal by using a total internal reflection geometry for the excitation by putting the sample on the face of a fused silica prism.[93,94] 2D resolution of about a micron was demonstrated. Two more recent application of 2D imaging are summarized below: nerve cell imaging[95,96] and monolayers on metals.[97–99]

Surface specific spectroscopy holds great promise for minimally invasive imaging in biological systems. Cellular structures form an interface between the largely aqueous solutions in the intercellular regions and the intracellular structures that are also largely bathed in an aqueous solution. Bringing the power of a vibrational spectroscopic technique into this environment would be a great advance. Sum frequency has so far not been used in biological environment, however, second harmonic generation has.[95,96] Attachment of fluorescent probes to biologically active molecules has reached a high art in recent years. These fluorescent probes are ideally suited to detection by second harmonic generation. Further, the SH response increases in the presence of an applied voltage (where the response becomes a third order susceptibility due to application of the voltage field). Although the increased response with field has not generally been characterized, for some fluorescent probes, the intensity increase is linear with applied voltage:[95]

$$\frac{I_{\mathrm{SHG}}(\phi) - I_{\mathrm{SHG}}(0)}{I_{\mathrm{SHG}}(0)} \propto \phi, \tag{36}$$

where $\phi$ is the applied potential. The sensitivity of $I_{SHG}$ to potential has been used to image changes in the membrane potential of neurons in response to applied stimuli. Nerves consist of a spine head separated from a dendrite by a slender spine neck. Specifically, the spine neck is too small to allow other techniques to probe the neck potential and thus unravel its role in signal transmission. The noninvasive nature of optical techniques avoids this spatial limitation. It was shown that the neck plays an important role in transmitting a potential by electrically isolating synapses. The demonstrated spatial resolution is submicron ($\sim$0.2 $\mu$m).[96] Further, optical techniques offer the potential to image multiple nerve structures simultaneously, making it possible to probe synergistic effects of multiple responses. This application of nonlinear spectroscopy is truly in its infancy.

Another successful approach to achieve 2D resolution of the surface is to access the enhanced nonresonant background from a metal substrate. As indicated in Eq. (15), the signal polarization adds onto or subtracts from the nonresonant polarization. Since the intensity is proportional to the square of the polarization, this effectively magnifies the effect of the signal. With this enhancement, the SFG signal can be imaged onto an array and still have sufficient signal to produce a spectrum. This strategy of using the nonresonant metal background to enhance the signal has been successfully implemented to image a self-assembled monolayer on gold,[99] CO on platinum,[98] and monolayers on mild steel.[100] Two-dimensional information is generated by dividing the 1 mm$^2$ SFG area into 10 $\mu$m×10 $\mu$m grids and extracting the spectrum from each grid spot. Specifically for CO on Pt, the grided spectra reveal that what was previously believed to be a very homogeneous surface, actually has distinctive regions: An average over the entire SF spot leads to the conclusion that CO sites at top sites as previously concluded.[101,102] Higher spatial resolution indicates that CO has two vibrational resonances: one at 2062 cm$^{-1}$ and the other at 2083 cm$^{-1}$.[98] The vibrational resonance of CO on Pt is known to be quite sensitive to the local environment,[103] so the presence of two resonances clearly demonstrates that the surface is inhomogeneous. Interpreting the two resonances awaits further experimental and theoretical developments, however, the existence of the two peaks suggests that even fairly well characterized surfaces can present surprises when the 2D resolution is improved. Indeed, further improvement of the 2D resolution is clearly needed for a large variety of surfaces.

## 5. Selected Results

### 5.1. *Ions at aqueous surfaces: The case for surface $H_3O^+$*

In this subsection, we review the experimental efforts aimed at testing the hypothesis that $H_3O^+$ partitions to the surface of an aqueous solution.[58,65,104] (Figure 12 shows a cartoon rendition of $H_3O^+$ on the aqueous surface.) To the extent that the experimental evidence points to $H_3O^+$ on the surface, that evidence comes from SHG and SFG experiments. Briefly, the theoretical results indicate a picture in which $H_3O^+$ sits on the surface due

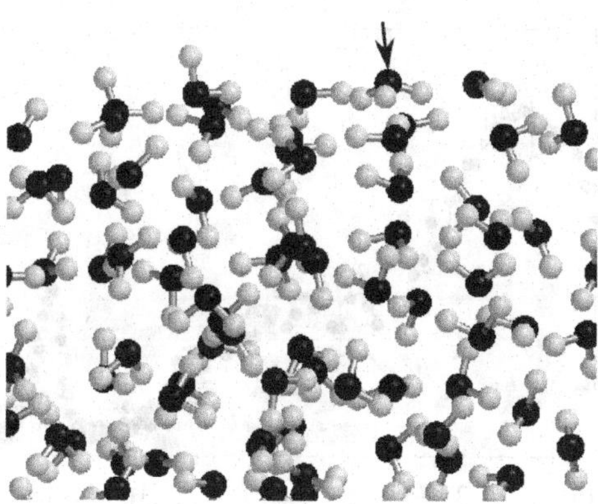

**Fig. 12.** Cartoon rendition of the aqueous surface with $H_3O^+$ in the top monolayer (marked with the arrow).

to the poor acceptor ability of the hydronium oxygen (formally, the oxygen carries a positive charge, hence does not attract a proton). Due to this poor acceptor ability, the coordination number of hydronium in solution is 3.6: lower than that of water.[105] Since water at the aqueous surface is forced to have a lower coordination than in the bulk, the hydronium ion is better accommodated there than is neutral water resulting in an enrichment of $H_3O^+$ at the surface. (Here we make no distinction between $H_3O^+$ and the more highly hydrated species: $H_5O_2^+$ etc.)

The first indication that the surface of acidic aqueous solutions are fundamentally different than water or nonacidic solutions was the near simultaneous reports[60,64] of a significant enhancement in the low frequency hydrogen-bonded peak in 0.01$x$ $H_2SO_4$ solutions compared to that of water. (See Fig. 13 for a typical sulfuric acid spectrum; 0.1$x$ HCl and $HNO_3$ show a similar enhancement, though not quite as dramatic.[62,63,106,107])

It may be tempting to attribute the enhanced intensity directly to the $H_3O^+$ ion. However, the $H_3O^+$ ion is expected to have a *very* broad resonance due to the labile nature of the proton.[108] Further, to be observed in the *ssp* spectrum, the $H_3O^+$ infrared dipole would have to be perpendicular to the interface,

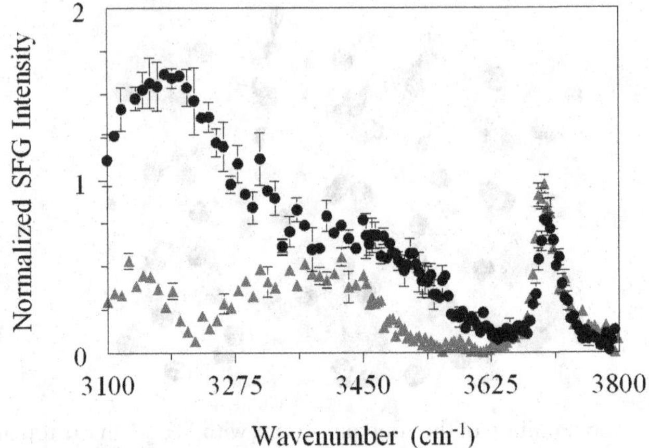


**Fig. 13.** The SFG spectrum of $0.1x$ $H_2SO_4$ (black circles) compared to that of neat water (gray triangles) in the *ssp* polarization combination shows a more than threefold enhancement of the strong hydrogen-bonded peak at about 3200 $cm^{-1}$.

i.e., the symmetric stretch. The symmetric stretch should be even more red shifted than the antisymmetric stretch, which should be at least as red shifted as the strongly hydrogen-bonded peak. The observation is that the peak intensity is enhanced, but not shifted from the strongly bonded peak. In addition, to produce the observed enhancement, the product of the surface enhancement and oscillator strength would have to be on the order of $10^4$–$10^5$ (at a minimum) that of hydrogen-bonded water for the enhancement to be directly due to $H_3O^+$.

Since the $H_3O^+$ is likely to be difficult to observe directly, recent efforts have focused on detecting enhancement of the acid counter ions at the interface, most recently that of $I^-$.[65,66,109,110] $I^-$ does not produce a vibrational spectrum, so SHG was used to detect its presence at the interface. In bulk solution, $I^-$ shows two charge transfer bands at 195 and 225 nm. These bands red shift at the aqueous surface[109] due to the reduced polarity at the surface. Observation of spectral enhancement at 200 nm for HI compared to that of NaI or KI leads to the conclusion that $H_3O^+$ partitions to the surface.[66] Interpretation of these $I^-$ experiments is somewhat controversial since SHG like SFG measures a convolution of the resonant and nonresonant signals. It is known that the nonresonant SH signal from water is not negligible. Furthermore, the nonresonant signal is affected by the orientation of water molecules in the surface. If that orientation changes, so does the nonresonant background. Ions in the interfacial layer influence the water molecules and affect their orientation. It is thus a challenge to separate the intensity enhancement that is due to orientation from enhancement that is directly due to the $I^-$ ions. Further, the connection between the $I^-$ concentration and that of $H_3O^+$ is not known *a prior*. Hence the SHG results provide, at best, indirect evidence of surface $H_3O^+$.

Very recently, on-the-fly molecular dynamics simulations were used to calculate the SFG spectrum from liquid water with a surface hydronium ion.[18] Although the agreement between the experimental spectrum and the calculated one is not quantitative (agreement is not expected to be quantitative since the simulation has no counter ion but the experiment always does) the simulation provides insight into the origin of the enhanced intensity. In both liquid water and ice, the strongest hydrogen-bonded peak is due to in-phase collective motions that involve vertically oriented OH bonds. (In ice, these vertical OH bonds bind the bilayers together.) With equal numbers of similarly oriented OH bonds pointing in and out of the bulk, the SF signal would be zero. However, the simulation indicates that those OH bonds that point into the solution have a tighter angular distribution than those pointing outward. The result of this difference in angular distribution, is a net inward pointing OH bond, the net orientation is responsible for the SFG signal. Addition of an $H_3O^+$ion, a positive charge, which partitions to the interface tightens the angular distribution of the inward facing OH bonds, enhancing the SF intensity.

Development of the picture of the location of ions at the aqueous interface, including the influence of the ions on the orientation and dynamics of water, is an on-going effort. Given the importance of aqueous solutions, particularly acidic solutions, it is likely that the next few years will see both increased activity in this area and a clearer picture.

### 5.2. *Interactions at nanostructured interfaces*

In this section, we review application of SFG in combination with small molecules to probe the surface of a nanoparticulate film, specifically that of $TiO_2$. Although a number of groups have applied SFG to study nanoparticle films based on Au nanoparticles,[111–114] there have been few reports of using SFG to probe the surface chemistry of semiconducting or insulating particles. This distinction is significant because metal nanoparticles are known to enhance the generated intensity. In addition, due to the image dipole developed by the metal in response to an oscillating dipole in the vicinity of the surface, usually only *p* polarized light generates a significant signal. Hence the polarization information that SFG can provide is missing.

Insulating or semiconducting nanoparticulate films raise an important question concerning polarization: How do these particles generate an SFG signal? If the particles are spherical and on the order of 1 nm diameter, there is insufficient phase shift for a nonzero SF intensity. Putting these spherical particles on a surface, however, breaks the spherical symmetry: if all sides are not accessible to an adsorbate, then a SF signal can be generated. Further, if the particle is nonspherical, then adsorption onto the surface could result in a large portion of the particle surface that is inaccessible: the other side can thus generate a signal. In the $TiO_2$ work, the approximately 2 nm particles are deposited on a $CaF_2$ substrate. The coated substrate is incorporated into a closed cell and the coated side exposed to a controlled gas phase. Several small molecules were used as chemical probes of the surface.[115–118] One of the probes is methanol.[116] Methanol adsorbs to the surface in two modes (Fig. 14): as molecular

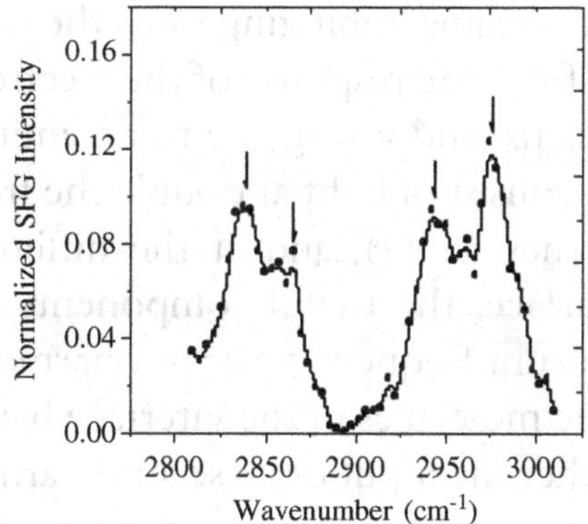


**Fig. 14.** SFG spectrum of 1.8 torr methanol on nanoparticulate $TiO_2$ at 20°C with *ssp* polarization. The four arrows mark (from low energy to high) the symmetric stretch of methoxy (2845 $cm^{-1}$), the symmetric stretch of molecular methanol (2873 $cm^{-1}$), the antisymmetric stretch of methoxy (2952 $cm^{-1}$), and the antisymmetric stretch of molecular methanol (2985 $cm^{-1}$).

methanol and dissociated as methoxy. Both of these species show a symmetric and an antisymmetric stretch mode (marked with arrows in Fig. 14). The symmetric and antisymmetric stretches of the $CH_3$ group are orthogonal resonances. Hence the polarization dependence of their intensities can be used to determine the orientation of the $CH_3$ symmetry axis with respect to the surface normal. The conclusion is that methoxy sits on the surface with the $C_3$ axis tilted 35° from the surface normal; molecular methanol sits on the surface tilted 70° from the normal.

In addition to indicating the orientation angle for methanol and methoxy on the surface, the polarization dependence is a definitive proof that the particles sit on the $CaF_2$ surface in such a way that only a portion of the crystal is accessible to the gas-phase probe molecules.

## 6. Summary

This paper presents a geometrical–optical tutorial on nonlinear optical signal generation: both SFG and SHG. The picture consists of the light beams impinging onto the surface, generating a polarization from the response of the electrons to the electric field of the beams, and giving rise to an emitted beam. The emitted beam can consist of light at double the frequency, SHG, at the sum frequency (SFG), and at the difference frequency (DFG). At the surface, the useful components are the second harmonic and the sum frequency. Second harmonic emission is enhanced when the molecules in the interface have an electronic resonance with either the input or its second harmonic. The sum frequency emission is enhanced when the interfacial molecules have a vibrational resonance with the impinging infrared radiation.

The surface response stems from the molecules in the interface: The relationship of the molecular orientation to the polarization characteristics of the excitation and emission is both complex and rich in information about the molecular orientation and concentration distribution at the interface. Unraveling this rich information makes SHG and SFG a challenge for those new to the field. Several molecular examples are given to aid in developing a picture of the nonlinear processes.

Recent developments use the information richness to determine an accurate orientation, including the direction of the surface dipole. This orientation and direction affect the molecular level interactions at the interface. These developments include using a reference, usually a nonresonant, signal to determine the dipole direction; exploiting orthogonal resonances to narrow the determined orientation; and using null angle techniques to more accurately determine the orientation. The null angle techniques use the inherent interference between polarizability elements, either by varying the visible input angle, VAN (visible angle null), or by varying the polarization angle, PAN (polarization angle null). Examples of both null angle techniques are given to illustrate the power of these methods.

Nonlinear spectroscopy is a fairly new technique in the array of methods for probing systems. Thus, it is still evolving. Current issues are surveyed. These include determination of the interfacial optical constants. The theoretical and experimental underpinnings for the choice of optical constants are just being developed. Beyond the dipole approximation, the bulk can also contribute to the signal. Experimental efforts are aimed at sorting out the bulk from the surface contribution to the signals. When the solvent is the object of the investigation, there can be strongly correlated motions that greatly enhance the signal. Current theoretical and experimental efforts are directed at determining these correlated modes and how they are affected by addition of solutes. The nonlinear spectroscopies are limited to a probe depth that is equal to the coherence length of the exciting beams: orders of magnitude longer than the molecular scale on which the effect of surface truncation is expected to bias the molecular orientation and therefore contribute to the signal. Much more experimental effort is needed to determine the practical probe depth.

Related to the probe depth is the issue of signal generation from nanoscale structures in solution. For structures as small as 55 nm, it has been demonstrated that the phase shift is sufficient to generate a signal, though not in the forward direction. Spectra from such structures are undoubtedly rich in information about the surface; a richness that has only barely been tapped. Similarly, the potential for ultrafast, time resolved studies has only barely been investigated. These time-resolved studies are expected to be particularly enlightening for the broad, hydrogen-bonded region of the water spectrum. Interpretation of spectra from the surface of water remains one of the biggest challenges to nonlinear spectroscopic methods and practioners. Despite the exquisite surface normal spatial resolution of nonlinear spectroscopy, extending that resolution into the surface plane has been accomplished only in rare cases. Two cases are discussed: nerve cell imaging and metal surface imaging.

Finally, two specific applications of nonlinear spectroscopy are discussed: the case for $H_3O^+$ at the aqueous surface and probing interactions at thin-films of insulating or semiconducting nanoparticles. Given the importance of aqueous surfaces to interstellar, atmospheric, biological, and electrochemical interfaces, the issue of ion distribution at the aqueous surface is very important. Arguably the most important, and certainly the most ubiquitous of these ions is the $H_3O^+$ ion. Even extremely pure water has an $H_3O^+$ concentration of $1 \times 10^{-7}$ M. If, as appears likely given the current evidence, $H_3O^+$ is on the very top monolayer capping dangling bonds, and $OH^-$ is further to the interior, the surface of water is slightly charged and has a strong dipole. Both the charge and the dipole affect the interaction of molecules impinging on the aqueous surface.

As nanoscale devices come into increasing use and their size get ever smaller, molecular scale interactions at the surface will become increasingly important. For a sufficiently small particle (depending on the dielectric constant of the particle) the size scale reaches a limit where there is no phase shift and no signal from these particles in solution. When a thin film of such particles is deposited on a substrate, there is a noncompensated side that can generate a signal. As with many nonlinear signals, the spectral signature is rich in information that has only begun to be tapped. In the case presented, that of $TiO_2$ nanoparticles, methanol is used as a molecular probe. It is found that these small particles have two adsorption modes for methanol: as molecular methanol and dissociated as methoxy. This demonstrates both that nonlinear spectroscopy is an excellent technique and that methanol is an information rich molecular probe.

The power of nonlinear spectroscopy has only just begun to be exploited to unravel information about interactions at a large variety of surfaces. The next several years promise tremendous development and an evolving understanding of surfaces.

**Acknowledgments**

Acknowledgement is made for partial support of the work discussed herein from NSF grants CHE-9816380, CHE0240172, CHE0613757, and PRF 37517-AC5. In addition, helpful input from Garth Simpson (Perdue University) and Steve Baldelli (University of Houston) is gratefully acknowledged. Finally, many hours of input were received from graduate students in the Shultz group — Henning Groenzin, Irene Li, Margaret Kuo, Nkengafeh Asong, Faith Dukes, Noelani Kamelamela, and Patrick Bisson — who read the manuscript with particular attention to portions that might keep the activation barrier high for those new to the field.

## Appendix A

### A.1 *Tensor product*

This appendix contains details of the tensor multiplication for the SFG output. The output polarization is

$$\tilde{P} = \tilde{\tilde{\tilde{\chi}}}^{(2)} : \tilde{E}_1 \tilde{E}_2, \tag{A.1}$$

where the tilde notation indicates the dimensionality of the tensors: $\tilde{P}$ is a first rank tensor, i.e., a vector, $\tilde{\tilde{\tilde{\chi}}}$ is a third order tensor. The intensity is then

$$I(\omega) = \frac{32\pi^3\omega^3 \sec\eta_{SF}^2}{c^3 n_1(\omega_{SF}) n_1(\omega_1) n_1(\omega_2)} |\tilde{E}(\omega_{SF}) \cdot \tilde{\tilde{\tilde{\chi}}}^{(2)} : \tilde{E}(\omega_{vis})\tilde{E}(\omega_{IR})|^2. \tag{A.2}$$

The dimensionally is as follows: $\tilde{E}$ are vectors — either 3 × 1 or 1 × 3, $\tilde{\tilde{\tilde{\chi}}}^{(2)}$ is a third rank tensor — a 3 × 3 × 3 array. Multiplication of the tensor by a vector results in a matrix, multiplication of the resulting matrix by a vector results in a vector. The final multiplication by the output vector, $\tilde{E}\,(\omega_{\mathrm{SF}})$ gives a scalar; the absolute value squared of which gives the intensity.

Walking through, $\tilde{\tilde{\tilde{\chi}}}^{(2)} : \tilde{E}_2$ is a 3 × 3 matrix with elements $(i = 1, 2, 3)$

$$\begin{pmatrix} \chi_{i11} & \chi_{i12} & \chi_{i13} \\ \chi_{i21} & \chi_{i22} & \chi_{i23} \\ \chi_{i31} & \chi_{i32} & \chi_{i33} \end{pmatrix} \begin{pmatrix} e_1(\omega_{\mathrm{IR}}) \\ e_2(\omega_{\mathrm{IR}}) \\ e_3(\omega_{\mathrm{IR}}) \end{pmatrix} = \begin{pmatrix} \chi_{i11}e_1(\omega_{\mathrm{IR}}) + \chi_{i12}e_2(\omega_{\mathrm{IR}}) + \chi_{i13}e_3(\omega_{\mathrm{IR}}) \\ \chi_{i21}e_1(\omega_{\mathrm{IR}}) + \chi_{i22}e_2(\omega_{\mathrm{IR}}) + \chi_{i23}e_3(\omega_{\mathrm{IR}}) \\ \chi_{i31}e_1(\omega_{\mathrm{IR}}) + \chi_{i32}e_2(\omega_{\mathrm{IR}}) + \chi_{i33}e_3(\omega_{\mathrm{IR}}) \end{pmatrix} \quad \text{(A.3)}$$

explicitly the 3 × 3 matrix:

$$\begin{pmatrix} \chi_{111}e_1(\omega_{\mathrm{IR}}) + \chi_{112}e_2(\omega_{\mathrm{IR}}) + \chi_{113}e_3(\omega_{\mathrm{IR}}) & \chi_{211}e_1(\omega_{\mathrm{IR}}) + \chi_{212}e_2(\omega_{\mathrm{IR}}) + \chi_{213}e_3(\omega_{\mathrm{IR}}) & \chi_{311}e_1(\omega_{\mathrm{IR}}) + \chi_{312}e_2(\omega_{\mathrm{IR}}) + \chi_{313}e_3(\omega_{\mathrm{IR}}) \\ \chi_{121}e_1(\omega_{\mathrm{IR}}) + \chi_{122}e_2(\omega_{\mathrm{IR}}) + \chi_{123}e_3(\omega_{\mathrm{IR}}) & \chi_{221}e_1(\omega_{\mathrm{IR}}) + \chi_{222}e_2(\omega_{\mathrm{IR}}) + \chi_{223}e_3(\omega_{\mathrm{IR}}) & \chi_{321}e_1(\omega_{\mathrm{IR}}) + \chi_{322}e_2(\omega_{\mathrm{IR}}) + \chi_{323}e_3(\omega_{\mathrm{IR}}) \\ \chi_{131}e_1(\omega_{\mathrm{IR}}) + \chi_{132}e_2(\omega_{\mathrm{IR}}) + \chi_{133}e_3(\omega_{\mathrm{IR}}) & \chi_{231}e_1(\omega_{\mathrm{IR}}) + \chi_{232}e_2(\omega_{\mathrm{IR}}) + \chi_{233}e_3(\omega_{\mathrm{IR}}) & \chi_{331}e_1(\omega_{\mathrm{IR}}) + \chi_{332}e_2(\omega_{\mathrm{IR}}) + \chi_{333}e_3(\omega_{\mathrm{IR}}) \end{pmatrix} \quad \text{(A.4)}$$

Dotting this matrix into the $\tilde{E}_1$ vector gives the vector:

$$
(e_1(\omega_{\mathrm{vis}})\; e_2(\omega_{\mathrm{vis}})\; e_3(\omega_{\mathrm{vis}}))\cdot
\begin{pmatrix}
\chi_{111}e_1(\omega_{\mathrm{IR}})+\chi_{112}e_2(\omega_{\mathrm{IR}})+\chi_{113}e_3(\omega_{\mathrm{IR}}) \\
\quad +\chi_{211}e_1(\omega_{\mathrm{IR}})+\chi_{212}e_2(\omega_{\mathrm{IR}})+\chi_{213}e_3(\omega_{\mathrm{IR}}) \\
\quad +\chi_{311}e_1(\omega_{\mathrm{IR}})+\chi_{312}e_2(\omega_{\mathrm{IR}})+\chi_{313}e_3(\omega_{\mathrm{IR}}) \\
\chi_{121}e_1(\omega_{\mathrm{IR}})+\chi_{122}e_2(\omega_{\mathrm{IR}})+\chi_{123}e_3(\omega_{\mathrm{IR}}) \\
\quad +\chi_{221}e_1(\omega_{\mathrm{IR}})+\chi_{222}e_2(\omega_{\mathrm{IR}})+\chi_{223}e_3(\omega_{\mathrm{IR}}) \\
\quad +\chi_{321}e_1(\omega_{\mathrm{IR}})+\chi_{322}e_2(\omega_{\mathrm{IR}})+\chi_{323}e_3(\omega_{\mathrm{IR}}) \\
\chi_{131}e_1(\omega_{\mathrm{IR}})+\chi_{132}e_2(\omega_{\mathrm{IR}})+\chi_{133}e_3(\omega_{\mathrm{IR}}) \\
\quad +\chi_{231}e_1(\omega_{\mathrm{IR}})+\chi_{232}e_2(\omega_{\mathrm{IR}})+\chi_{233}e_3(\omega_{\mathrm{IR}}) \\
\quad +\chi_{331}e_1(\omega_{\mathrm{IR}})+\chi_{332}e_2(\omega_{\mathrm{IR}})+\chi_{333}e_3(\omega_{\mathrm{IR}})
\end{pmatrix},
\tag{A.5}
$$

$$
\begin{pmatrix}
[\chi_{111}e_1(\omega_{\mathrm{IR}})+\chi_{112}e_2(\omega_{\mathrm{IR}})+\chi_{113}e_3(\omega_{\mathrm{IR}})]e_1(\omega_{\mathrm{vis}}) \\
\quad +[\chi_{121}e_1(\omega_{\mathrm{IR}})+\chi_{122}e_2(\omega_{\mathrm{IR}})+\chi_{123}e_3(\omega_{\mathrm{IR}})]e_2(\omega_{\mathrm{vis}}) \\
\quad +[\chi_{131}e_1(\omega_{\mathrm{IR}})+\chi_{132}e_2(\omega_{\mathrm{IR}})+\chi_{133}e_3(\omega_{\mathrm{IR}})]e_3(\omega_{\mathrm{vis}}) \\
[\chi_{211}e_1(\omega_{\mathrm{IR}})+\chi_{212}e_2(\omega_{\mathrm{IR}})+\chi_{213}e_3(\omega_{\mathrm{IR}})]e_1(\omega_{\mathrm{vis}}) \\
\quad +[\chi_{221}e_1(\omega_{\mathrm{IR}})+\chi_{222}e_2(\omega_{\mathrm{IR}})+\chi_{223}e_3(\omega_{\mathrm{IR}})]e_2(\omega_{\mathrm{vis}}) \\
\quad +[\chi_{231}e_1(\omega_{\mathrm{IR}})+\chi_{232}e_2(\omega_{\mathrm{IR}})+\chi_{233}e_3(\omega_{\mathrm{IR}})]e_3(\omega_{\mathrm{vis}}) \\
[\chi_{311}e_1(\omega_{\mathrm{IR}})+\chi_{312}e_2(\omega_{\mathrm{IR}})+\chi_{313}e_3(\omega_{\mathrm{IR}})]e_1(\omega_{\mathrm{vis}}) \\
\quad +[\chi_{321}e_1(\omega_{\mathrm{IR}})+\chi_{322}e_2(\omega_{\mathrm{IR}})+\chi_{323}e_3(\omega_{\mathrm{IR}})]e_2(\omega_{\mathrm{vis}}) \\
\quad +[\chi_{331}e_1(\omega_{\mathrm{IR}})+\chi_{332}e_2(\omega_{\mathrm{IR}})+\chi_{333}e_3(\omega_{\mathrm{IR}})]e_3(\omega_{\mathrm{vis}})
\end{pmatrix}.
\tag{A.6}
$$

The vector in Eq. (A.6) is the generated polarization. The output electric field is obtained from this polarization via the nonlinear optical factors, Table 3. The numerator in the $L(\mathrm{SF})$ factors arise as follows. The $Y$ component of the polarization is perpendicular to the propagation direction, so $e_2(\mathrm{SF})$ is determined by the transmission factor. The parallel component is more involved since the generated polarization is, in general, not orthogonal to the propagation direction (Eqs. (3) and (4)). The generated polarization can be projected onto the output field direction as follows: the vectors $\cos\eta_{r,\mathrm{SF}} P_X$ and $\sin\eta_{r,\mathrm{SF}} P_Z$ are both parallel to the output field direction. The sum of these thus generates the output field.

$$\begin{pmatrix} P(\omega_{\mathrm{SF}})\cos\eta_{r,\mathrm{SF}}\cos\Theta \\ P(\omega_{\mathrm{SF}})\sin\Theta \\ P(\omega_{\mathrm{SF}})\sin\eta_{r,\mathrm{SF}}\cos\Theta \end{pmatrix}, \tag{A.7}$$

where $\Theta$ is the SF polarization angle. The output intensity is proportional to the square of

$$\begin{pmatrix} \{[\chi_{111}e_1(\omega_{\mathrm{IR}})+\chi_{112}e_2(\omega_{\mathrm{IR}})+\chi_{113}e_3(\omega_{\mathrm{IR}})]e_1(\omega_{\mathrm{vis}}) \\ \quad +[\chi_{121}e_1(\omega_{\mathrm{IR}})+\chi_{122}e_2(\omega_{\mathrm{IR}})+\chi_{123}e_3(\omega_{\mathrm{IR}})]e_2(\omega_{\mathrm{vis}}) \\ \quad +[\chi_{131}e_1(\omega_{\mathrm{IR}})+\chi_{132}e_2(\omega_{\mathrm{IR}})+\chi_{133}e_3(\omega_{\mathrm{IR}})]e_3(\omega_{\mathrm{vis}}) \\ \quad \times P(\omega_{\mathrm{SF}})\cos\eta_{r,\mathrm{SF}}\cos\Theta \\ \\ \{[\chi_{211}e_1(\omega_{\mathrm{IR}})+\chi_{212}e_2(\omega_{\mathrm{IR}})+\chi_{213}e_3(\omega_{\mathrm{IR}})]e_1(\omega_{\mathrm{vis}}) \\ \quad +[\chi_{221}e_1(\omega_{\mathrm{IR}})+\chi_{222}e_2(\omega_{\mathrm{IR}})+\chi_{223}e_3(\omega_{\mathrm{IR}})]e_2(\omega_{\mathrm{vis}}) \\ \quad +[\chi_{231}e_1(\omega_{\mathrm{IR}})+\chi_{232}e_2(\omega_{\mathrm{IR}})+\chi_{233}e_3(\omega_{\mathrm{IR}})]e_3(\omega_{\mathrm{vis}})\} \\ \quad \times P(\omega_{\mathrm{SF}})\sin\Theta \\ \\ \{[\chi_{311}e_1(\omega_{\mathrm{IR}})+\chi_{312}e_2(\omega_{\mathrm{IR}})+\chi_{313}e_3(\omega_{\mathrm{IR}})]e_1(\omega_{\mathrm{vis}})\} \\ \quad +[\chi_{321}e_1(\omega_{\mathrm{IR}})+\chi_{322}e_2(\omega_{\mathrm{IR}})+\chi_{323}e_3(\omega_{\mathrm{IR}})]e_2(\omega_{\mathrm{vis}}) \\ \quad +[\chi_{331}e_1(\omega_{\mathrm{IR}})+\chi_{332}e_2(\omega_{\mathrm{IR}})+\chi_{333}e_3(\omega_{\mathrm{IR}})]e_3(\omega_{\mathrm{vis}})\} \\ \quad \times P(\omega_{\mathrm{SF}})\sin\eta_{r,\mathrm{SF}}\cos\Theta \end{pmatrix}. \tag{A.8}$$

The output electric field is simplified by eliminating all elements of the second order hyperpolaizability that are zero:

$$\begin{pmatrix} \{\chi_{113}e_3(\omega_{\mathrm{IR}})e_1(\omega_{\mathrm{vis}}) + \chi_{131}e_1(\omega_{\mathrm{IR}})e_3(\omega_{\mathrm{vis}}) \\ \quad \times P(\omega_{\mathrm{SF}}) \cos\eta_{\mathrm{r,SF}} \cos\Theta \\ \{\chi_{223}e_3(\omega_{\mathrm{IR}})e_2(\omega_{\mathrm{vis}}) + \chi_{232}e_2(\omega_{\mathrm{IR}})e_3(\omega_{\mathrm{vis}})\} \\ \quad \times P(\omega_{\mathrm{SF}}) \sin\Theta \\ \{\chi_{311}e_1(\omega_{\mathrm{IR}})e_1(\omega_{\mathrm{vis}})\} + \chi_{322}e_2(\omega_{\mathrm{IR}})e_2(\omega_{\mathrm{vis}}) \\ \quad + \chi_{333}e_3(\omega_{\mathrm{IR}})e_3(\omega_{\mathrm{vis}})\} \times P(\omega_{\mathrm{SF}}) \sin\eta_{\mathrm{r,SF}} \cos\Theta \end{pmatrix}. \tag{A.9}$$

Null angle SFG fixes the infrared polarization at $p$, so $e_2(\omega_{\mathrm{IR}}) = 0$, and the output becomes

$$\begin{pmatrix} \{\chi_{113}e_3(\omega_{\mathrm{IR}})e_1(\omega_{\mathrm{vis}}) + \chi_{131}e_1(\omega_{\mathrm{IR}})e_3(\omega_{\mathrm{vis}}) \\ \quad \times P(\omega_{\mathrm{SF}}) \cos\eta_{\mathrm{r,SF}} \cos\Theta \\ \chi_{223}e_3(\omega_{\mathrm{IR}})e_2(\omega_{\mathrm{vis}}) \times P(\omega_{\mathrm{SF}}) \sin\Theta \\ \{\chi_{311}e_1(\omega_{\mathrm{IR}})e_1(\omega_{\mathrm{vis}})\} + \chi_{333}e_3(\omega_{\mathrm{IR}})e_3(\omega_{\mathrm{vis}})\} \\ \quad \times P(\omega_{\mathrm{SF}}) \sin\eta_{\mathrm{r,SF}} \cos\Theta \end{pmatrix}. \tag{A.10}$$

Case I: If the visible is also $p$-polarized, then $e_2(\omega_{\mathrm{vis}}) = 0$

$$\begin{pmatrix} \{\chi_{113}e_3(\omega_{\mathrm{IR}})e_1(\omega_{\mathrm{vis}}) + \chi_{131}e_1(\omega_{\mathrm{IR}})e_3(\omega_{\mathrm{vis}}) \\ \quad \times P(\omega_{\mathrm{SF}}) \cos\eta_{\mathrm{r,SF}} \cos\Theta \\ 0 \\ \{\chi_{311}e_1(\omega_{\mathrm{IR}})e_1(\omega_{\mathrm{vis}})\} + \chi_{333}e_3(\omega_{\mathrm{IR}})e_3(\omega_{\mathrm{vis}})\} \\ \quad \times P(\omega_{\mathrm{SF}}) \sin\eta_{\mathrm{r,SF}} \cos\Theta \end{pmatrix}. \tag{A.11}$$

Only the $p$-polarized sum frequency survives. That is, this is *ppp*.

Case II: If the visible is $s$-polarized, then only $e_2(\omega_{\mathrm{vis}}) \neq 0$

$$\{\chi_{223}e_3(\omega_{\mathrm{IR}})e_2(\omega_{\mathrm{vis}})\} \times P_2(\omega_{\mathrm{SF}}). \tag{A.12}$$

Only the $s$ term survives for SF, this is *ssp*.

Returning to (A.9), the *ssp* intensity is related to (note: $1 = x$, $2 = y$, $3 = z$),

$$\left|\chi_{223} e_3(\omega_{\text{IR}}) e_2(\omega_{\text{vis}}) \times P(\omega_{\text{SF}}) \sin\Theta\right|^2 \tag{A.13}$$

The Fresnel factors convert from $e_i(\omega_{\text{beam}})$ to $e_{beam}$ (aside from overall multiplicative factors), e.g., $e_1(\omega_{\text{vis}}) = K_x^{\text{vis}} e_{\text{vis}}^p$, $e_2(\omega_{\text{vis}}) = K_Y^{\text{vis}} e_{\text{vis}}^s$, and $\mathcal{P}(\omega_{\text{SF}}) \sin\Theta = L_Y$

$$L_Y \chi_{YYZ} K_Z^{\text{IR}} e_{\text{IR}}^p K_Y^{\text{vis}} e_{\text{vis}}^s = A_{ssp} \tag{A.14}$$

as in Eq. (8).

The *sps* intensity is related to

$$\left|\chi_{232} e_2(\omega_{\text{IR}}) e_3(\omega_{\text{vis}}) \times P(\omega_{\text{SF}}) \sin\Theta\right|^2 \tag{A.15}$$

and converting yields,

$$\chi_{YZY} K_Y^{\text{IR}} e_{\text{IR}}^s K_Z^{\text{vis}} e_{\text{vis}}^p L_Y = A_{ssp} \tag{A.16}$$

as in Eq. (8). Finally, the *ppp* intensity is related to

$$\left|\begin{array}{l} \{\chi_{113} e_3(\omega_{\text{IR}}) e_1(\omega_{\text{vis}}) + \chi_{131} e_1(\omega_{\text{IR}}) e_3(\omega_{\text{vis}})\} \\ \quad \times P(\omega_{SF}) \cos\eta_{\text{r,SF}} \cos\Theta \\ \quad + \{\chi_{311} e_1(\omega_{\text{IR}}) e_1(\omega_{\text{vis}}) + \chi_{333} e_3(\omega_{\text{IR}}) e_3(\omega_{\text{vis}})\} \\ \quad \times P(\omega_{\text{SF}}) \sin\eta_{\text{r,SF}} \cos\Theta \end{array}\right|^2 . \tag{A.17}$$

$$\begin{aligned} A_{ppp} \equiv{} & \cos\eta_{\text{r,SF}} L_X \chi_{XXZ} K_X^{\text{vis}} K_Z^{\text{IR}} e_{\text{vis}}^p e_{\text{IR}}^p \\ & + \cos\eta_{\text{r,SF}} L_X \chi_{XZX} K_Z^{\text{vis}} K_X^{\text{IR}} e_{\text{vis}}^p e_{\text{IR}}^p \\ & + \sin\eta_{\text{r,SF}} L_Z \chi_{ZXX} K_X^{\text{vis}} K_X^{\text{IR}} e_{\text{vis}}^p e_{\text{IR}}^p \\ & + \sin\eta_{\text{r,SF}} L_Z \chi_{ZZZ} K_Z^{\text{vis}} K_Z^{\text{IR}} e_{\text{vis}}^p e_{\text{IR}}^p \end{aligned} \tag{A.18}$$

Eq. (A.18) indicates that changing the visible angle alters the balance between $e_1(\omega_{\text{vis}})$ and $e_3(\omega_{\text{vis}})$.

## A.2 *Null angle*

The produced polarization vector is

$$\begin{pmatrix} [\chi_{113}e_3(\omega_{\mathrm{IR}})]e_1(\omega_{\mathrm{vis}}) + [\chi_{131}e_1(\omega_{\mathrm{IR}})]e_3(\omega_{\mathrm{vis}}) \\ [\chi_{223}e_3(\omega_{\mathrm{IR}})]e_2(\omega_{\mathrm{vis}})] \\ [\chi_{311}e_1(\omega_{\mathrm{IR}})]e_1(\omega_{\mathrm{vis}})] + [\chi_{333}e_3(\omega_{\mathrm{IR}})]e_3(\omega_{\mathrm{vis}}) \end{pmatrix} \quad \text{(A.19)}$$

In the null angle technique, the IR is fixed at $p$, the visible at a polarization of 45°, so the visible is 0.707 $p$-polarized, 0.707 $s$-polarized. The analysis polarizer is rotated from −90° to +90° with a null generated when the SF polarization is orthogonal to the analyzer.

$$\begin{aligned} &\{0.707\chi_{XXZ}K_Z^{\mathrm{IR}}e_{\mathrm{IR}}^{p}K_X^{\mathrm{vis}}e_{\mathrm{vis}} + 0.707\chi_{XZX}K_X^{\mathrm{IR}}e_{\mathrm{IR}}^{p}K_Z^{\mathrm{vis}}e_{\mathrm{vis}}\} \\ &\quad \times L_X\cos\Theta + 0.707\chi_{YYZ}K_Z^{\mathrm{IR}}e_{\mathrm{IR}}^{p}K_Y^{\mathrm{vis}}e_{\mathrm{vis}}^{s} \times L_Y\sin\Theta \\ &\quad + \{0.707\chi_{ZXX}K_X^{\mathrm{IR}}e_{\mathrm{IR}}^{p}K_X^{\mathrm{vis}}e_{\mathrm{vis}}^{p} + 0.707\chi_{ZZZ}K_Z^{\mathrm{IR}}e_{\mathrm{IR}}^{p}K_Z^{\mathrm{vis}}e_{\mathrm{vis}}^{p}\} \\ &\quad \times L_Z\cos\Theta \qquad \text{(A.20)} \end{aligned}$$

and with the canceling terms removed:

$$\begin{aligned} &0.707\chi_{XXZ}K_Z^{\mathrm{IR}}e_{\mathrm{IR}}^{p}K_X^{\mathrm{vis}}e_{\mathrm{vis}} \times L_X\cos\Theta \\ &\quad + 0.707\chi_{YYZ}K_Z^{\mathrm{IR}}e_{\mathrm{IR}}^{p}K_Y^{\mathrm{vis}}e_{\mathrm{vis}} \times L_Y\sin\Theta \\ &\quad + 0.707\chi_{ZZZ}K_Z^{\mathrm{IR}}e_{\mathrm{IR}}^{p}e_{\mathrm{vis}} \times L_Z\cos\Theta. \qquad \text{(A.21)} \end{aligned}$$

### References


1. Y. R. Shen, *Ann. Rev. Phys. Chem.* **40**, 327 (1989).
2. M. J. Shultz, C. Schnitzer, D. Simonelli and S. Baldelli, *Int. Rev. Phys. Chem.* **19**, 123 (2000).
3. K. B. Eisenthal, *Chem. Rev.* **96**, 1343 (1996).
4. R. Corn and D. Higgins, *Chem. Rev.* **94**, 107 (1994).
5. S. Gopalakrishnan, D. Liu, H. C. Allen, M. Kuo and M. J. Shultz, *Chem. Rev.* **106**, 1155 (2006).
6. K. B. Eisenthal, *Chem. Rev.* **106**, 1462 (2006).
7. H.-F. Wang, W. Gan, R. Lu, Y. Rao and B.-H. Wu, *Int. Rev. Phys. Chem.* **24**, 191 (2005).

8. M. J. Shultz, S. Baldelli, C. Schnitzer and D. Simonelli, *J. Phys. Chem. B* **106**, 5315 (2002).
9. C. D. Bain, P. B. Davies, T. H. Ong and R. N. Ward, *Langmuir* 7, 1563 (1991).
10. H. Yamamoto, N. Akamatsu, A. Wada and C. Hirose, *J. Elect. Spec. Relat. Phenom.* **64/65**, 507 (1993).
11. N. Akamatsu, K. Domen, C. Hirose and H. Yamamoto, *J. Phys. Chem.* **97**, 10064 (1993).
12. C. Hirose, N. Akamatsu and K. Domen, *Appl. Spec.* **46**, 1051 (1992).
13. C. Hirose, N. Akamatsu and K. Domen, *J. Chem. Phys.* **96**, 997 (1992).
14. N. Akamatsu, K. Domen and C. Hirose, *Appl. Spec.* **46**, 1051 (1992).
15. Y. Rao, Y.-S. Tao and H.-F. Wang, *J. Chem. Phys.* **119**, 5226 (2003).
16. N. Bloembergen and P. S. Pershan, *Phys. Rev.* **128**, 606 (1962).
17. V. Buch and J. P. Devlin, *J. Chem. Phys.* **110**, 3437 (1999).
18. V. Buch, G. Richmond, H. Groenzin, I. Li and M. J. Shultz, , *J. Chem. Phys.* **127**, 204710 (2007).
19. M. Born and E. Wolf, *Principles of Optics: Electromagnetic Theory of Propagation, Interference and Diffraction of Light* (Cambridge University Press, 1999), pp. 38–48.
20. X. Wei, S.-C. Hong, X. Zhuang, T. Goto and Y. R. Shen, *Phys. Rev. E* **62**, 5160 (2000).
21. E. Hecht, *Optics* (Addison Wesley Longman, Inc., 1998).
22. R. Superfine, J. Y. Huang and Y. R. Shen, *Opt. Lett.* **15**, 1276 (1990).
23. R. Superfine, J. Y. Huang and Y. R. Shen, *Chem. Phys. Lett.* **172**, 303 (1990).
24. K. Kemnitz, K. Bhattacharyya, J. M. Hicks, G. R. Pinto, K. B. Eisenthal and T. F. Heinz, *Chem. Phys. Lett.* **131**, 285 (1986).
25. D. Simonelli and M. J. Shultz, *J. Chem. Phys.* **112**, 6804 (2000).
26. Y. Rao, M. Comstock and K. B. Eisenthal, *J. Phys. Chem. B* **110**, 1727 (2006).
27. H. Chen, W. Gan, B.-H. Wu, D. Wu, Z. Zhang and H.-F. Wang, *Chem. Phys. Lett.* **408**, 284 (2005).
28. W. Gan, B.-H. Wu, H. Chen, Y. Guo and H.-F. Wang, *Chem. Phys. Lett.* **406**, 467 (2005).
29. A. G. Lambert, P. B. Davies and D. J. Neivandt, *Appl. Spec. Rev.* **40**, 103 (2005).
30. D. E. Gragson and G. L. Richmond, *J. Phys. Chem. B* **102**, 3847 (1998).
31. V. Ostroverkhov, G. A. Waychunas and Y. R. Shen, *Phys. Rev. Lett.* **94**, 046102 (2005).
32. G. J. Simpson, C. A. Dailey, R. M. Plocinik, A. J. Moad, M. A. Polizzi and R. M. Everly, *Anal. Chem.* 77, 215 (2005).
33. X. D. Zhu, H. Suhr and Y. R. Shen, *Phys. Rev. B* **35**, 3047 (1987).
34. P. Guyot-Sionnest and Y. R. Shen, *Phys. Rev. B* **35**, 4420 (1987).
35. V. Pouthier, C. Ramseyer and C. Girardet, *J. Chem. Phys.* **108**, 6502 (1998).
36. C. Girardet, J. Humbert and P. N. M. Hoang, *Chem. Phys.* **230**, 67 (1998).

37. X. Zhuang, P. B. Miranda, D. Kim and Y. R. Shen, *Phys. Rev. B* **59**, 12632 (1999).
38. T. F. Heinz, H. W. K. Tom and Y. R. Shen, *Phys. Rev. A* **28**, 1883 (1983).
39. T. F. Heinz, C. K. Chen, D. Ricard and Y. R. Shen, *Phys. Rev. Lett.* **48**, 478 (1982).
40. M. B. Feller, W. Chen and Y. R. Shen, *Phys. Rev. A* **43**, 6778 (1991).
41. V. Mizrahi and J. E. Sipe, *J. Opt. Soc. Am. B* **5**, 660 (1988).
42. X. Zhuang, D. Wilk, L. Marrucci and Y. R. Shen, *Phys. Rev. Lett.* **75**, 2144 (1995).
43. A. J. Fordyce, W. J. Bullock, A. J. Timson, S. Haslan, R. D. Spencer-Smith, A. Alexander and J. G. Frey, *Mol. Phys.* **99**, 677 (2001).
44. J. A. Ekhoff and K. L. Rowlen, *Anal. Chem.* **74**, 5954 (2002).
45. Y. A. Grudzkov and E. N. Parmon, *J. Chem. Soc. Faraday Trans.* **89**, 4017 (1993).
46. G. J. Simpson, S. Westerbuhr and K. L. Rowlen, *Anal. Chem.* **72**, 887 (1993).
47. G. Cnossen, K. E. Drabe and D. A. Wiersma, *J. Chem. Phys.* **97**, 4512 (1992).
48. H. Hsiung, G. R. Meredith, H. Vanherzeele, R. Popovitz-Biro, E. Shavit and M. Lahav, *Chem. Phys. Lett.* **164**, 539 (1989).
49. R. M. Plocinik and G. L. Simpson, *Anal. Chim. Acta* **496**, 133 (2003).
50. X. Wei, S.-C. Hong, A. I. Lvovsky, H. Held and Y. R. Shen, *J. Phys. Chem. B* **104**, 3349 (2000).
51. V. Buch, S. Bauerecker, J. P. Devlin, U. Buck and J. K. Kazimirski, *Int. Rev. Phys. Chem.* **23**, 375 (2004).
52. F. Huisken, A. Kulcke, C. Laush and J. Lisy, *J. Chem. Phys.* **95**, 3924 (1991).
53. A. Morita and J. T. Hynes, *J. Phys. Chem.* **B 106**, 673 (2002).
54. T. Ishiyama and A. Morita, *J. Phys. Chem. C* **111**, 738 (2007).
55. T. Ishiyama and A. Morita, *J. Phys. Chem. C* **111**, 721 (2007).
56. A. Perry, C. Neipert, C. Ridley, B. Space and P. B. Moore, *Phys. Rev.* **71**, 050601:1 (2005).
57. D. Walker, D. Hore and G. L. Richmond, *J. Phys. Chem. B* **110**, 20451 (2006).
58. M. Mucha, T. Frigato, L. M. Levering, H. C. Allen, D. J. Tobias, L. X. Dang and P. Jungwirth, *J. Phys. Chem. B* **109**, 7617 (2005).
59. E. Brown, M. Mucha, P. Jungwirth and D. Tobias, *J. Phys. Chem. B* **109**, 7934 (2005).
60. S. Baldelli, C. Schnitzer, M. J. Shultz and D. Campbell, *J. Phys. Chem. B* **101**, 10435 (1997).
61. S. Baldelli, C. Schnitzer, M. J. Shultz and D. J. Campbell, *Chem. Phys. Lett.* **287**, 143 (1998).
62. C. Schnitzer, S. Baldelli, D. J. Campbell and M. J. Shultz, *J. Phys. Chem. A* **103**, 6383 (1999).
63. S. Baldelli, C. Schnitzer and M. J. Shultz, *Chem. Phys. Lett.* **302**, 157 (1999).
64. C. Radüge, V. Pflumio and Y. R. Shen, *Chem. Phys. Lett.* **274**, 140 (1997).

65. M. K. Petersen, S. S. Iyengar, T. J. F. Day and G. A. Voth, *J. Phys. Chem. B* **108**, 14804 (2004).
66. P. B. Petersen and R. J. Saykally, *J. Phys. Chem. B* **109**, 7976 (2005).
67. G. Ritzhaupt and J. P. Devlin, *J. Phys. Chem.* **95**, 90 (1991).
68. C. Aliaga and S. Baldelli, *J. Phys. Chem. B* **110**, 18481 (2006).
69. X. Zhang and R. A. Walker, *Langmuir* **17**, 4486 (2001).
70. W. Steel, F. Damkaci, R. Nolan and R. Walker, *J. Am. Chem. Soc.* **124**, 4824 (2002).
71. X. Zhang, H. Steel and R. A. Walker, *J. Phys. Chem. B* **107**, 3829 (2003).
72. S. Baldelli, *J. Phys. Chem. B* **109**, 13049 (2005).
73. C. Laurence, P. Nicolet and M. T. Dalati, *J. Phys. Chem.* **98**, 5807 (1994).
74. D. K. Lambert, *J. Chem Phys.* **89**, 3847 (1988).
75. D. K. Lambert, *Electrochim. Acta* **41**, 623 (1996).
76. S. Chang, L. H. Leung and M. J. Weaver, *J. Phys. Chem* **93**, 5341 (1989).
77. S. Wantanabe, J. Inukai and M. Ito, *Surf. Sci.* **293**, 1 (1993).
78. M. J. Weaver and X. Gao, *Ann. Rev. Phys. Chem.* **44**, 459 (1993).
79. Y. Liu, J. I. Dadap, D. Zimdars and K. B. Eisenthal, *J. Phys. Chem. B* **103**, 2480 (1999).
80. J. I. Dadap, J. Shan, K. B. Eisenthal and T. F. Heinz, *Phys. Rev. Lett.* **83**, 4045 (1999).
81. J. Shan, J. I. Dadap, I. Stiopkin, G. A. Reider and T. F. Heinz, *Phys. Rev. A* **73**, 023819:1 (2006).
82. S. Roke, W. Roeterdink, J. Wijnhoven, A. Petukhov, A. Kleyn and M. Bonn, *Phys. Rev. Lett.* **90**, 258302:1 (2003).
83. S. Roke, M. Bonn and A. V. Petukhov, *Phys. Rev. B* **70**, 115106:1 (2004).
84. Y. Pavlyukh and A. Hübner, *Phys. Rev. B* **70**, 245434:1 (2004).
85. S.-H. Jen and H.-L. Dai, *J. Phys. Chem. B* **110**, 23000 (2006).
86. M. Morgenthaler and S. Meech, *J. Phys. Chem.* **100**, 3323 (1996).
87. D. Zimdars, J. I. Dadap, K. B. Eisenthal and T. F. Heinz, *J. Phys. Chem. B* **103**, 3425 (1999).
88. D. Zimdars, J. Dadap, K. Eisenthal and T. Heinz, *Chem. Phys. Lett.* **301**, 112 (1999).
89. A. V. Benderskii and K. B. Eisenthal, *J. Phys. Chem. B* **104**, 11723 (2000).
90. A. Benderskii and K. Eisenthal, *J. Phys. Chem. A* **106**, 7482 (2002).
91. J. McGuire and Y. R. Shen, *Science* **313**, 1945 (2006).
92. A. Bordenyuk, H. Jayathilake and A. Benderskii, *J. Phys. Chem.* **109**, 15941 (2005).
93. M. Flörscheimer, C. Brillert and H. Fucks, *Langmuir* **15**, 5437 (1999).
94. M. Flörsheimer, C. Brillert and H. Fuchs, *Mat. Sci. Eng.* **8–9**, 335 (1999).
95. M. Nuriya, J. Jiang, B. Nemet, K. Eisenthal and R. Yuste, *Proc. Nat. Acad. Sci. US* **103**, 786 (2006).

96. R. Araya, J. Jiang, K. Eisenthal and R. Yuste, *Proc. Nat. Acad. Sci. US* **103**, 17961 (2006).
97. K. Kuhnke, F. M. P. Hoffmann, X. C. Wu, A. M. Bittner and K. Kern, *Appl. Phys. Lett.* **83**, 3830 (2003).
98. K. Cimatu and S. Baldelli, *J. Am. Chem. Soc.* **128**, 16016 (2006).
99. K. Cimatu and S. Baldelli, *J. Phys. Chem. B* **110**, 1807 (2006).
100. S. Baldelli, H. Zhang and C. Romero, Investigation of alkanethiol monolayers on mild steel with sum frequency generation and electrochemistry, 2006, Personal Communication.
101. I. Villegas and M. J. Weaver, *J. Am. Chem. Soc.* **118**, 458 (1996).
102. P. Hollins and J. Pritchard, *Prog. Surf. Sci.* **19**, 275 (1985).
103. T. Iwasita and F. C. Nart, *Prog. Surf. Sci.* **55**, 271 (1997).
104. T. James and D. J. Wales, *J. Chem. Phys.* **122**, 134306 (2005).
105. T. J. F. Day, A. V. Soudackov, M. Uma, U. W. Schmitt and G. A. Voth, *J. Chem. Phys.* **117**, 5839 (2002).
106. C. Schnitzer, S. Baldelli and M. J. Shultz, *J. Phys. Chem. B* **104**, 585 (2000).
107. S. Baldelli, D. Campbell, C. Schnitzer and M. J. Shultz, *J. Phys. Chem. B* **103**, 2789 (1999).
108. H. D. Downing and D. Williams, *J. Phys. Chem.* **80**, 1640 (1976).
109. P. B. Petersen, J. C. Johnson, K. P. Knutsen and R. J. Saykally, *Chem. Phys. Lett.* **397**, 46 (2004).
110. P. B. Petersen and R. J. Saykally, *J. Phys. Chem. B* **110**, 14060 (2006).
111. C. Humbert, B. Busson, J. P. Abid, C. Six, H. H. Girault and A. Tadjeddine, *Electrochim. Acta* **50**, 3101 (2005).
112. G. Rupprechter, *Phys. Chem. Chem. Phys.* **3**, 4621 (2001).
113. J. Holman, S. Ye, D. Neivandt and P. Davies, *J. Am. Chem. Soc.* **126**, 14322 (2004).
114. T. Kawai, D. Keivandt and P. B. Davies, *J. Am. Chem. Soc.* **122**, 12031 (2000).
115. C.-Y. Wang, H. Groenzin and M. J. Shultz, *Langmuir* **19**, 7330 (2003).
116. C.-Y. Wang, H. Groenzin and M. J. Shultz, *J. Phys. Chem. B* **108**, 265 (2004).
117. C.-Y. Wang, H. Groenzin and M. J. Shultz, *J. Am. Chem. Soc.* **126**, 8094 (2004).
118. C.-Y. Wang, H. Groenzin and M. J. Shultz, *J. Am. Chem. Soc.* **127**, 9736 (2005).

Chapter 5

# Propagation and Intramolecular Coupling Effects in the Four-Wave Mixing Spectroscopy

José Luis Paz

*Departamento de Química, Universidad Simón Bolívar, Apartado 89000, Caracas 1080[a], Venezuela*
*jlpaz@usb.ve*

In this work, we have studied the changes in the Four-Wave Mixing spectroscopic signal response in the frequency space by propagation of the electromagnetic fields and intramolecular coupling effects. Firstly, in a propagation treatment at all orders in the pump field and first order in the probe and signal fields, we obtain a description of the Four-Wave Mixing signal. Secondly, for the inclusion of the intramolecular effects, the model is a generalization of the conventional two-level approach in that it includes a simplified description of the molecular structure. The Born–Oppenheimer electronic energy curves for this simplified molecular model consist of two intercrossing harmonic oscillators potential with different force constants. The transition and permanent dipole moments are critical quantities for this analysis in the frequency space of the local macroscopic polarization in the adiabatic representation when the rotating wave approximation is not considered.

## 1. Introduction

During the past three decades a number of powerful methods have been developed to study the linear and nonlinear light–matter interaction. Different studies are reported to date in the literature, where: (a) the semiclassical optical conventional Bloch equations (OCBE) are used as the starting point to characterize these radiative interaction; (b) the non-radiative terms are described phenomenologically introducing the longitudinal ($T_1$) and transversal ($T_2$) relaxation times; and (c) the state of the system is described in terms of the reduce density matrix after all necessary contractions over the variables associated with the environment (considered as a thermal bath) are performed. The resulting equations, widely used to describe the effect of the fields acting on a statistical ensemble, are then used to study in detail the inhomogeneous or homogeneous character of the resulting spectral lines. Homogeneous contributions to the inhomogeneous broadening of the resulting bands, and measurements of the longitudinal and the transversal relaxation times of the electronic transitions involved, are of significant importance in spectroscopy. In particular, Four-Wave mixing spectroscopy is one of the most useful methods that can be used to obtain such data.[1–4] In most realistic situations, the radiative effects are certainly not the only source and, usually, they are not the dominant source of relaxation fluctuations and incoherences. Depending on the source of the fluctuations, different modifications to $T_1$ and $T_2$ may occur. In these cases, molecular collisions induce frequency shifts and, due to the nature of the collisions, itself may lead to a stochastic modulation of the transition frequency. In studies realized by Wodkiewicz[5] it has been shown that the OCBE in the presence of a random field lead to a set of multiplicative stochastic differential equations, for which, the exact solutions can be analytically obtained. In this case, if the relaxation mechanism involves changes in the molecular transition frequency of the system and if these frequency shifts persist following a molecule–field perturbation interaction, then the equations governing the time development of the Bloch-vector are integro-differential in nature and the OCBE are no longer valid.

The interaction of an electromagnetic wave with the nonlinear medium can produce various effects. Sum or difference frequency, parametric amplification, and wave mixing are some examples of nonlinear processes with multiple applications. The

Four-Wave Mixing (FWM) process is of special interest in our research work and this technique has been employed in the determination of $T_1$ (associated with the relaxation mechanisms of the populations to the Boltzmann type equilibrium regime) and $T_2$ (associated with mechanisms of lost of coherence on the system in absence of the incident field) in semiconductors and organic dyes.[1,2] These relaxation times are relatively short for most dyes and they are accurately determined from nonlinear spectroscopy, e.g., using the FWM. Many phenomena are noticeable when electromagnetic fields are allowed to pass through a given optical path. In connection, some methods using Wei-Norman algebraic procedures for the study of field propagation through homogeneous system (where an analogy is established between the evolution system operator and the so-called optic propagation matrices) have been developed.[6] Authors like Boyd, worked out propagation light-field models by assuming constant intensity valued for the pump beam along the optical path.[7] Since this approximation is not valid in general. Reif *et al.*[8] take into account the homogeneous spectral line and consider the case in which all the fields propagate through the medium for the FWM signal. Theoretical studies most often use models consisting of systems of two or more levels to describe the active system interacting with the fields, while the environment is modeled through a thermal bath.[9,10] In the context of these general theories, Bavli and Band[9,11] used harmonic generation technique for the sum (SFG) and the difference of frequency (DFG) to study new peaks appearing at $\omega_1 = 0$ and $\omega_2 = \omega_0/5$ outside the resonant region. They also studied the absorption and dispersion profiles of a two-level system with a nonzero permanent dipole moment interacting with two low intensity beams at frequencies given by $\omega_1 = \omega_2 = \omega_0/2$. Similarly, Lavoine *et al.*[10] studied the effect of including permanent dipole moments in the degenerate FWM technique, finding also peaks outside the resonance region at frequency $\omega_1 = \omega_2 = \omega_3 = \omega_0/2$. In the works of Bessega *et al.*[12] are studied the characteristics of the resonances emerging as a consequence of incorporating a permanent dipole moment in the non-degenerate Four-Wave Mixing (nd-FWM) technique, which are investigated in the framework of a local homogeneous-linewidth third order perturbation model. In this work, the results show that the effects are the permanent dipole moments on the nonlinear signal topology are a consequence of neglecting the rotating-wave approximation (RWA). In the work of Paz *et al.*[13] they applied the nd-FWM technique, including the permanent molecular dipole moments, to describe the response of a two-level system irradiated by classical electromagnetic fields treated to third order. The importance of nonzero permanent dipole moments for linear and nonlinear optical properties has been shown previously by Shimoda and Shimizu,[14] Leasure *et al.*,[15] Thomas and Meath,[16] Dick and Hohlneicher,[17] Meath and Power,[18] Kmetic and Meath,[19] Band *et al.*,[20] Bavli and Band,[9,11,21] Bavli *et al.*[22] and García-Sucre *et al.*[23,24]

In the literature there exist theoretical and experimental works related to FWM processes. Some of this work considers a classical electromagnetic field interacting with a two-level system without specifying the internal characteristic of the levels involved. However, it is possible to conceive of a two-level system with vibrational internal structure; in other words, it is possible to conceive of it as vibrational states belonging to potential energy curves in different electronic states. This potential curves can be coupled, and this is related to the intramolecular coupling phenomenon. In this way, one could build all the variables related to their vibrational structure into the model.

In this work, we studied the effects of the intramolecular coupling on the nd-FWM nonlinear signal response and to present the reduced polarization expression that permit the different resonances in a FWM spectrum to be characterized. In order to obtain this objective, we have considered a homogeneous resonance frequency distribution of molecular two-level system interacting simultaneously with a thermal reservoir and with classical electromagnetic fields treated as plane waves. The two-level involved are considered as different electronic states described by crosses one-dimensional harmonic potentials, each electronic level having only one fundamental vibrational level associated to its potential curve. This model will be used presently to simulate an aqueous solution of organic Malachite Green dye. The radiation–matter interaction is described by the coupling to the non-zero permanent dipole moments of the uncoupled states, which has been proved to contribute considerably to the photonic processes occurring in the nonresonant regions of the spectrum.[9,11,17,22,25] In the literature there are many works related to the intramolecular effect. In some works related to time resolved signals for complex dye molecules in solution, the system is conceived as few-level density-matrix models with phenomenological dephasing constants, stochastic modulation of the levels, harmonic-oscillator models, or microscopic models of non Born–Oppenheimer dynamics on multidimensional excited-state potential energy surface.[26] Moreover, the intramolecular coupling is of great importance in various physical and chemical processes, for instance in the study of vibronic coupling in weak transition probabilities forbidden by symmetry in absorption and emission,[27] in the study of relaxation rates of internal conversion,[28] in femtosecond spectroscopy (employing a simple case of an intersection of harmonic diabatic potential energy surfaces),[29] in the study of the optical absorption band shape of dimmers,[30] and in the study of resonances in a scattering process.[31]

The vibronic coupling resulting by coupling the nuclear and electronic motions in a molecule is of great importance to some physical chemistry processes.[28] The curve crossing problems have received more attention and they have found applications in various fields of physics, chemistry, and biology.[32]

Finally, for this reviewer work, we have presented a formalism to study the optical properties on a two-level molecular system interacting with the electromagnetic field, but considering explicitly the presence of the solvent. We consider a model where molecular collisions induce frequency shifts that lead to a modulation of the transition frequency.

## 2. Four-Wave Mixing Spectroscopy

### 2.1. *Study and characterization of FWM signal in the frequency space*

The forward FWM has been shown to be very useful in matter–light interaction studies and as a measuring method for population and memory phase relaxation times. The experiments involving two incident laser beams on an optically active solute dissolved in water. In this technique, two beam laser are used (field of pumping of frequency $\omega_1$, and probe field of frequency $\omega_2$) and they are focused inside a resonant medium one, frequently generating a signal at frequency $\omega_3 = 2\omega_1 - \omega_2$, and the vector of wave $\vec{k}_3 \approx 2\vec{k}_1 - \vec{k}_2$. Figure 1 shows two FWM processes with the energy levels diagrams, respectively.

We describe the time-dependent process of interaction of a molecule with a total external field and with a heat reservoir using the Liouville–Von Newmann formalism in the semiclassical approximation, consisting of neglecting the quantization of the electromagnetic field. The Liouville equation may be written as $d\rho(t)/dt = -\frac{i}{\hbar}[H, \rho(t)]$, where $\rho(t)$ is the reduced density matrix, $H$ is the total Hamiltonian given by $H = H_{\mathrm{m}} + H_{\mathrm{mf}} + H_{\mathrm{mR}}$. In this Hamiltonian, $H_{\mathrm{m}}$ corresponds to the isolated molecule, $H_{\mathrm{mf}}$ gives the interaction between the molecule and the external field, and $H_{\mathrm{mR}}$ is associated with the interaction between the molecule and the heat reservoir. We treat here the matrix representation of Liouville equation in the uncoupled

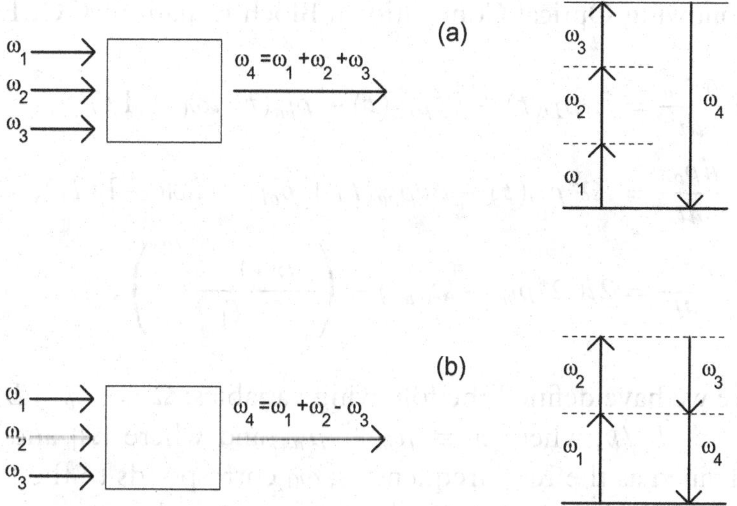


**Fig. 1.** Four-Wave Mixing processes.

basis of the molecular states $|a\rangle$ and $|b\rangle$, where $H_{\mathrm{m}}$ is diagonal. In the dipole approximation, $H_{\mathrm{mf}}$ is given by $H_{\mathrm{mf}} = -\vec{\mu} \cdot \vec{E}$. In this work, we have considered the matrix associated with the dipole moment as $\underline{\mu} = \begin{bmatrix} \mu_{aa} & \mu_{ab} \\ \mu_{ba} & \mu_{bb} \end{bmatrix}$, where $\mu_{aa}$ and $\mu_{bb}$ represent the permanent dipole moments associated with the states $|a\rangle$ and $|b\rangle$, respectively, and $\mu_{ab}$ and $\mu_{ba}$ are the transition dipole moments between the states $|a\rangle$ and $|b\rangle$. Finally, the contribution of the commutator $[H_{\mathrm{mR}}, \rho(t)]$ to the time evolution of the density matrix $\rho(t)$ may be represented following the usual phenomenological description of the relaxation mechanisms as $[H_{\mathrm{mR}}, \rho(t)]_{ab} = -i\hbar(\rho_{ab} - \rho_{ab}^{(0)})\gamma_{ab}$.[33] When $a = b$, $\rho_{aa}^{(0)}$ corresponds to the fraction of molecules in the state $|a\rangle$ for an equilibrium distribution and $\rho_{ab}^{(0)}$ is taken as zero for $a \neq b$. Similarly, we define $\gamma_{jj} = 1/T_1(j = a, b)$ and $\gamma_{ij} = \gamma_{ji} = 1/T_2(i, j = a, b)$, where $T_1$ and $T_2$ are the longitudinal and transversal relaxation times, respectively. With all the above considerations in mind, the evolution of the density matrix of a two-level system in a thermal reservoir and in the presence of electromagnetic field is given by the following Optical Conventional Bloch Equation (OCBE):

$$\frac{d\rho_{ba}}{dt} = i\Omega\rho_D(t) + i\Phi\rho_{ba}(t) - \rho_{ba}(t)(i\omega_0 + 1/T_2), \quad (1a)$$

$$\frac{d\rho_{ab}}{dt} = i\Omega^*\rho_D(t) - i\Phi\rho_{ba}(t) + \rho_{ab}(t)(i\omega_0 - 1/T_2), \quad (1b)$$

$$\frac{d\rho_D}{dt} = 2i(\Omega^*\rho_{ba} - \Omega\rho_{ab}) - \left(\frac{\rho_D(t) - \rho_D^{(0)}}{T_1}\right), \quad (1c)$$

where we have defined the following variables: $\Omega = (\vec{\mu}_{ba} \cdot \vec{E})/\hbar$; $\phi = (d \cdot E)/\hbar$, where $d \equiv \mu_{bb} - \mu_{aa}$ and where $|\Omega|$ and $|\Phi|$ are defined as the Rabí frequencies; $\omega_0$ corresponds to the Bohr frequency of the two-level molecular system and $\rho_D(t) = \rho_{aa} - \rho_{bb}$ is the population difference, while that $\rho_D^{(0)}$ is the equilibrium value of the population difference in the absence of radiation. The field $E(r,t)$ corresponds to the amplitude of the total external electromagnetic field, given by $E = E_1 + E_2$, where $E_1$ is the pump field, and $E_2$ is the probe field, and where we have defined $E_m = E_{0m}\cos[\vec{k}_m \cdot \vec{r} - \omega_m t + \phi_m] = E_m(\omega_m)\exp(-i\omega_m t) + c.c.$ with m = 1, 2, and $E_m(\omega_m) = (E_{0m}/2)\exp[i(\vec{k}_m \cdot \vec{r}) + \phi_m]$ represent the Fourier component that oscillates at frequency $\omega_m$.

Equations (1) above are Fourier transformed to the frequency domain to obtain

$$\rho_{ba}(\omega_3) = \nu_3\Omega_2^* + \beta_3\Omega_3, \quad (2a)$$

$$\rho_{ab}(-\omega_2) = \beta_2^*\Omega_2^* + \nu_2^*\Omega_3, \quad (2b)$$

$$\rho_{ba}(\omega_1) = \beta_1\Omega_1, \quad (2c)$$

where

$$\nu_3 = \frac{-2i\Omega_1^2\left(\frac{1}{D_2^*} + \frac{1}{D_1}\right)}{D_3\left[\Gamma + 2|\Omega_1|^2\left(\frac{1}{D_2^*} + \frac{1}{D_3}\right)\right]}\rho_D^{dc};$$

$$\beta_3 = i\frac{2|\Omega_1|^2\left(\frac{1}{D_2^*} - \frac{1}{D_1^*}\right) + \Gamma}{D_3\left[\Gamma + 2|\Omega_1|^2\left(\frac{1}{D_2^*} + \frac{1}{D_3}\right)\right]}\rho_D^{dc}$$

$$\beta_2^* = i\frac{2|\Omega_1|^2\left(\frac{1}{D_1} - \frac{1}{D_3}\right) - \Gamma}{D_2^*\left[\Gamma + 2|\Omega_1|^2\left(\frac{1}{D_2^*} + \frac{1}{D_3}\right)\right]}\rho_D^{dc};$$

$$\nu_2^* = \frac{2i\Omega_1^{*2}\left(\frac{1}{D_3} + \frac{1}{D_1^*}\right)}{D_2^*\left[\Gamma + 2|\Omega_1|^2\left(\frac{1}{D_2^*} + \frac{1}{D_3}\right)\right]}\rho_D^{dc};$$

$$\beta_1 = \frac{i}{D_1}\rho_D^{dc},$$

where the population difference component that oscillates at zero frequency is given by

$$\rho_D^{dc} = \frac{|D_1|^2 T_2^2}{|D_1|^2 T_2^2 + 4S}\rho_D^0, \quad \text{with}$$

$$D_j = \frac{1}{T_2} + i(\omega_0 - \omega_j) \ (j = 1, 2, 3).$$

$\Gamma = (1/T_1) - i\Delta$, with $\Delta = \omega_1 - \omega_2$; $S$ corresponds to the saturation parameter (representing a measure of the radiation–matter intensity coupling) defined by $S = |\Omega_1|^2 T_1 T_2$ with $\Omega_k = \vec{\mu}_{ba} \cdot \vec{E}(\omega_j)/\hbar$. In the present model, we have used the semiclassical approximation, so our formulation does not include spontaneous relaxation channels. We have neglected the absorption of the generated signal along the optical path. Taking into account all the above considerations, the intensity in the local case is given by $I(\omega_3) = (c\varepsilon/8\pi)|P(t)|^2$, where $P(t)$ is the macroscopic polarization, expressed as $P(t) = N\langle\vec{\mu}\rangle = NTr(\rho(t)\vec{\mu})$.

In the case of inhomogeneous broadening, the frequency component of the induced macroscopic polarization is calculated according to

$$P(\omega_j) = N \int_{-\infty}^{\infty} d\omega_0 L(\omega_0) \langle \rho_{ba}(\omega_j) \vec{\mu}_{ab} \rangle, \tag{3}$$

where $N$ is the solute concentration, the angular bracket representing an average over the distribution of molecular orientations, and $L(\omega_0)$ corresponds to an inhomogeneous Lorentzian distribution of the resonance frequencies, given by $L(\omega_0) = \frac{1}{\pi}\left[\frac{\gamma}{\gamma^2+(\omega_0-\omega_c)^2}\right]$. In this equation, $\gamma$ represents the "half width at half maximum" (HWHM), and $\omega_c$ corresponds to the centre of the inhomogeneously broadened absorption line. By considering all orders of perturbation in the pump field, but only first order in the probe and signal fields, and by using the scalar, the rotating wave, and the steady state approximations, the following relationships were derived for the inhomogeneous complex polarization at the pump, probe and signal frequencies $P^{\text{inh}}(\omega_j)$. The following expressions were obtained

$$P(\omega_3) = \chi_{ef}^{(3)}(\omega_3)E^2(\omega_1)E^*(-\omega_2) + \tilde{\chi}_{ef}^{(3)}(\omega_3)E(\omega_3) + \chi_{ef}^{(1)}(\omega_3)E(\omega_3) + \chi^{(\text{sv})}(\omega_3)E(\omega_3), \tag{4a}$$

$$P(\omega_2) = \chi_{ef}^{(3)}(\omega_2)E^2(\omega_1)E^*(-\omega_3) + \tilde{\chi}_{ef}^{(3)}(\omega_2)E(\omega_2) + \chi_{ef}^{(1)}(\omega_2)E(\omega_2) + \chi^{(\text{sv})}(\omega_2)E(\omega_2), \tag{4b}$$

$$P(\omega_1) = \chi_{ef}^{(1)}(\omega_1)E(\omega_1) + \chi^{(\text{sv})}(\omega_1)E(\omega_1), \tag{4c}$$

where $\chi^{(\text{sv})}(\omega_j)$ and $\chi_{ef}^{(\text{m,inh})}(\omega_j)$ represent the solvent and scalar inhomogeneous complex susceptibilities at frequency $\omega_j$, respectively. The superscript m represents the minimum order required for the contribution to be considered. The expressions for the different susceptibilities are given in this model by:

$$\chi_{ef}^{(3,\mathrm{inh})}(\omega_k) = \frac{4N\gamma\mu_{ab}^4\rho_D^0}{\hbar^3 T_2\left(\Gamma\delta_{k,3}+\Gamma^*\delta_{k,2}\right)}\left[\frac{2}{T_2}\pm(-i\Delta)\right] \times \left\{\frac{1+\gamma T_2 - i\left(\omega_c-\omega_1\right)T_2}{A_\pm}+\frac{1+\sqrt{1+4S}}{B_\pm} + \frac{1-iT_2\left(\pm a+bi\right)}{C_\pm}\right\}, \tag{5}$$

$$\tilde{\chi}_{ef}^{(3,\mathrm{inh})}(\omega_k) = (\pm)\frac{-4iN\gamma\mu_{ab}^4\rho_D^0\Delta}{\hbar^3 T_2\left(\Gamma\delta_{k,3}+\Gamma^*\delta_{k,2}\right)} \times \left\{\frac{1-\gamma T_2 + i\left(\omega_c-\omega_1\right)T_2}{A_\pm}+\frac{1-\sqrt{1+4S}}{B_\pm} + \frac{1+iT_2\left(\pm a+bi\right)}{C_\pm}\right\} E\left(\omega_1\right)E^*\left(-\omega_1\right), \tag{6}$$

$$\chi_{ef}^{(1,\mathrm{inh})}(\omega_k) = \frac{-2N\gamma\mu_{ab}^2\rho_D^0}{\hbar T_2^3}\left\{\frac{\left[1+\gamma T_2 - i(\omega_c-(\omega_1-n\Delta))T_2\right]\times\left[1+T_2^2(\omega_c-\omega_1+i\gamma)^2\right]}{A_\pm} - \frac{4S\left[\pm(-)\,iT_2\Delta+1+\sqrt{1+4S}\right]}{B_\pm} + \frac{\left[1\pm(-iT_2)(\Delta+a\pm bi)\right]\times\left[1+T_2^2(\pm a+bi)^2\right]}{C_\pm}\right\}, \tag{7}$$

where the triads of symbols $(n, k, \pm)$ correspond to $(1, 3, +)$ for the signal field, and to $(-1,\ 2, -)$ for the probe field. Here, the susceptibility for the pump field:

$$\chi_{ef}^{(1)}(\omega_1) = \frac{-2N\gamma\mu_{ab}^2\rho_D^0}{\hbar T_2}\left\{\frac{1+\gamma T_2 - i(\omega_c-\omega_1)T_2}{A_0} + \frac{1+\sqrt{1+4S}}{B_0}\right\}, \tag{8}$$

where the coefficients are written as

$$A_0 = 2i\gamma\left[(\omega_c-\omega_1+i\gamma)^2 + \frac{1+4S}{T_2^2}\right];$$

$$B_0 = 2i\left[\left(\omega_1-\omega_c+i\frac{\sqrt{1+4S}}{T_2}\right)^2+\gamma^2\right]\left[\frac{\sqrt{1+4S}}{T_2}\right]$$

$$A_\pm = A_0\left[(\omega_c-\omega_1+i\gamma)^2-(\pm a+bi)^2\right];$$

$$B_\pm = -B_0\left[\frac{1+4S}{T_2^2}+(\pm a+bi)^2\right]$$

$$C_\pm = 2\left[(\omega_1-\omega_c\pm a+bi)^2+\gamma^2\right]\left[\frac{1+4S}{T_2^2}+(\pm a+bi)^2\right] \times [\pm a+bi] \quad \text{with} \quad a+bi=\sqrt{\omega_1^2-\omega_q^2},$$

$$\omega_q^2 = \omega_2\omega_3+\frac{1}{T_2^2}+\frac{4S(1+T_1T_2\Delta^2)}{T_2^2(1+T_1^2\Delta^2)} + \left(\frac{2i\Delta}{T_2}\right)\left[\frac{2S(T_1-T_2)}{T_2(1+T_1^2\Delta^2)}-1\right] = \omega_{qR}^2+i\omega_{qI}^2,$$

where

$$a = -\frac{\omega_{qI}^2}{\sqrt{2}\sqrt{\omega_{qR}^2-\omega_1^2+\sqrt{\left(\omega_{qR}^2-\omega_1^2\right)^2+\left(\omega_{qI}^2\right)^2}}},$$

$$b = \frac{1}{\sqrt{2}}\sqrt{\omega_{qR}^2 - \omega_1^2 + \sqrt{\left(\omega_{qR}^2 - \omega_1^2\right)^2 + \left(\omega_{qI}^2\right)^2}}.$$

Considering the permanent dipole moment in the model, the frequency component of the induced polarization is then calculated for a homogeneous linewidth according to:

$$P(\omega_j) = N\langle\rho_{aa}(\omega_j)\vec{\mu}_{aa} + \rho_{bb}(\omega_j)\vec{\mu}_{bb} + \rho_{ab}(\omega_j)\vec{\mu}_{ba} + \rho_{ba}(\omega_j)\vec{\mu}_{ab}\rangle_\Theta. \tag{9}$$

Introducing the perturbation expansion of the density matrix $\rho(t)$ (second order in the pump field and first order in the probe) and by using the steady-state approximation in calculating the different components (coherence and population) of the density matrix at the optical frequency of interest $\omega_3$, we are led to the following expression for the induced polarization:

$$\begin{aligned}
P(\omega_3) = N\Bigg\{\rho_D^{(0)}\mu_{ba}\Bigg[\frac{1}{(D_3^+)^*}\Bigg[2\Omega_1^2\Omega_2^*\left(\frac{1}{\Gamma}\left(\frac{1}{D_2^+} + \frac{1}{(D_1^+)^*}\right.\right.\\
\left.\left. + \frac{1}{D_1^-} + \frac{1}{(D_2^-)^*}\right) + \frac{1}{\lambda}\left(\frac{1}{D_1^-} + \frac{1}{(D_2^-)^*}\right)\right)\\
+ \frac{\Phi_1^2\Omega_2^*}{(D_2^{;-})^*(D_\Delta^+)^*} + \frac{\Phi_1\Phi_2^*\Omega_1}{(D_1^{;+})^*}\left(\frac{1}{(D_{2\omega_1}^{;+})^*(D_\Delta^+)^*}\right)\Bigg]\\
- \frac{1}{D_3^-}\Bigg[2\Omega_1^2\Omega_2^*\left(\frac{1}{\Gamma}\left(\frac{1}{D_2^+} + \frac{1}{(D_1^+)^*} + \frac{1}{D_1^-} + \frac{1}{(D_2^-)^*}\right)\right.\\
\left. + \frac{1}{\lambda}\left(\frac{1}{D_1^-} + \frac{1}{(D_2^-)^*}\right)\right) + \frac{\Phi_1^2\Omega_2^*}{(D_2^{;+})(D_\Delta^-)}\\
+ \frac{\Phi_1\Phi_2^*\Omega_1}{(D_1^{;-})}\left(\frac{1}{(D_{2\omega_1}^{;-})^*(D_\Delta^-)}\right)\Bigg]
\end{aligned}$$

$$+\frac{i\rho_D^{(0)}\vec{d}}{\beta}\left[\Omega_1^2\Omega_2^*\left(\frac{1}{(D_1^-)(D_\Delta^-)}-\frac{1}{(D_1^+)^*(D_\Delta^+)^*}\right)\right.$$

$$+\Omega_1\Omega_2^*\Phi_1\left(\frac{1}{(D_{2\omega_1}^-)(D_1^-)}-\frac{1}{(D_2^-)^*(D_\Delta^+)^*}\right.$$

$$\left.\left.+\frac{1}{(D_2^+)(D_\Delta^-)}-\frac{1}{(D_{2\omega_1}^+)^*(D_1^+)^*}\right)\right]\Bigg]\Bigg\}, \tag{10}$$

where

$$D_j^\pm=\frac{1}{T_2}+i\left(\omega_0\pm\omega_j\right)\quad j=1,2,3;$$

$$D_\Delta^\pm=\frac{1}{T_2}+i\left(\omega_0\pm\Delta\right);$$

$$D_{2\omega_1}^\pm=\frac{1}{T_2}+i\left(\omega_0\pm 2\omega_1\right);\quad \lambda=\frac{1}{T_1}-2i\omega_1;$$

$$\beta=\frac{1}{T_1}-i\omega_3;$$

$$\Gamma=\frac{1}{T_1}-i\Delta;\ \Omega_j=\frac{\vec{\mu}_{ba}\cdot\vec{E}_j}{\hbar};\quad \Phi_j=\frac{\vec{d}\cdot\vec{E}_j}{\hbar};$$

$$\Delta=\omega_1-\omega_2.$$

It will be shown below that the signal's intensity will be the same in the regions of optical resonance whether we make these differences in permanent dipole moments zero or not. In such a case, the macroscopic nonlinear polarization is given by

$$P(\omega_3)=-\frac{2iN}{\hbar^3}\vec{\mu}_{ab}\frac{\left(\vec{\mu}_{ba}\cdot\vec{E}_1\right)^2\left(\vec{\mu}_{ba}\cdot\vec{E}_2^*\right)\left(D_1^-+(D_2^-)^*\right)}{\Gamma D_3^- D_1^-(D_2^-)^*}\rho_D^{(0)}, \tag{11}$$

where the contribution of the antiresonant terms to the induced polarization are not taken into account.

## 2.2. *Effects of solute concentration, field intensity, and spectral inhomogeneous broadening on FWM*

### 2.2.1. *Propagation effects*

The propagation in the present model was studied starting from the Maxwell equation given by:[34]

$$\nabla^2 \mathbf{E} - \frac{1}{c^2}\left(\frac{\partial^2 \mathbf{E}}{\partial t^2}\right) = \frac{4\pi}{c^2}\left(\frac{\partial^2 \mathbf{P}^{\text{inh}}}{\partial t^2}\right) \tag{12}$$

and using the slowly varying envelope approximation

$$\left|\frac{d^2 \tilde{E}_j}{dz^2}\right| \ll \left|2ik_j \frac{d\tilde{E}_j}{dz}\right| \tag{13}$$

to obtain the following equations for the propagation of the fields along the $z$ direction:

$$\frac{d\tilde{E}_1}{dz} = -\alpha_1^{\text{inh}} \tilde{E}_1 \quad \text{(pump)}, \tag{14a}$$

$$\frac{d\tilde{E}_2}{dz} = -\alpha_2^{\text{inh}} \tilde{E}_2 + \xi_2^{\text{inh}} \tilde{E}_3^* \exp\left(i\Delta k_z z\right) \quad \text{(probe)}, \tag{14b}$$

$$\frac{d\tilde{E}_3}{dz} = -\alpha_3^{\text{inh}} \tilde{E}_3 + \xi_3^{\text{inh}} \tilde{E}_2^* \exp\left(i\Delta k_z z\right) \quad \text{(signal)}. \tag{14c}$$

$\tilde{E}_j (j = 1, 2, 3)$ represents the envelopes of the fields given by $\vec{E}_j = (1/2)E_{0j} \exp(i\Phi_j)$; $\alpha_j^{\text{inh}}$ represents the inhomogeneous (inh) nonlinear absorption coefficient of the material medium at frequency $\omega_j$ in the presence of the pump beam, given by

$$\alpha_1 = \frac{2\pi\omega_1}{c\eta_1} \text{Im}\chi_{ef}^{(1)}(\omega_1) \tag{15a}$$

and

$$\alpha_j = \frac{2\pi\omega_j}{c\eta_j} \text{Im}\left[\tilde{\chi}_{ef}^{(3)}(\omega_j) + \chi_{ef}^{(1)}(\omega_j)\right] \quad (j = 2, 3), \tag{15b}$$

the coefficients $\xi_j^{\text{inh}}$ correspond to the inhomogeneous coupling parameters between the probe and signal field, given by $\xi_j^{\text{inh}} = i\frac{2\pi\omega_j}{c\eta_j}\chi_{ef}^{(3)}(\omega_j)\,\tilde{E}_1^2$ for $j = 2, 3$ (probe and signal beams), and $\Delta k_z$ represents the $z$ component of the propagation-vector mismatch:

$$\Delta k_z = 2k_{1_z} - k_{2_z} - k_{3_z} = \frac{\omega}{c}\left[2\eta_1 - (\eta_2 + \eta_3)\cos\theta\right], \quad (15c)$$

where $\theta$ is the separation angle between the probe and pump beams. The refractive index in this model is defined by

$$\eta_1 = \sqrt{\eta_0^2 + 4\pi \mathrm{Re}\chi_{ef}^{(1)}(\omega_1)} \quad \text{(pump field)} \quad (16a)$$

and $\eta_j$ represent the probe and signal beams inhomogeneous refractive indexes given by

$$\eta_j = \sqrt{\eta_0^2 + 4\pi \mathrm{Re}\left[\tilde{\chi}_{ef}^{(3)}(\omega_j) + \chi_{ef}^{(1)}(\omega_j)\right]} \quad (16b)$$

for the probe $j = 2$, and signal fields $j = 3$. The solvent refractive index is given by

$$\eta_0 = \sqrt{1 + 4\pi\ \mathrm{Re}\ \chi^{\mathrm{sv}}(\omega_j)} \quad (j = 1, 2, 3). \quad (16c)$$

The above equations show that the refractive indexes are the result of adding the real part of the coherent and incoherent contributions of the corresponding nonlinear susceptibility, where the incoherent component takes into account the reduction in the relative population due to the saturative pump field, and the presence of the coherent component, which involves interference between the weak and strong fields, is due to population oscillations at the detuning frequency $\Delta$ between the pump and probe fields. We also observe here, as shown in a previous study,[35] all possible energy gains of the probe and the signal beams (treated at first order) have to come from the absorptive interaction of the pump field (treated at all order) with the molecular system. Decoupling (14), we obtain:

$$\frac{d^2\tilde{E}_3}{dz^2} + \Psi_1^{(\mathrm{S,inh})}(z)\frac{d\tilde{E}_3}{dz} + \Psi_2^{(\mathrm{S,inh})}(z)\tilde{E}_3 = 0, \qquad (17)$$

where the inhomogeneous coefficients are given by

$$\Psi_1^{(\mathrm{S,inh})}(z) = \alpha_2^{(\mathrm{S,inh})} + \alpha_3^{(\mathrm{S,inh})} - \frac{d\ln\xi_3^{(\mathrm{S,inh})}}{dz} - i\Delta k_z - iz\frac{d\Delta k_z}{dz} \qquad (18a)$$

and

$$\Psi_2^{(\mathrm{S,inh})}(z) = \alpha_2^{(\mathrm{S,inh})}\alpha_3^{(\mathrm{S,inh})} - \xi_2^{(\mathrm{S,inh})*}\xi_3^{(\mathrm{S,inh})} + \frac{d\alpha_3^{(\mathrm{S,inh})}}{dz} - \alpha_3^{(\mathrm{S,inh})}\left(\frac{d\ln\xi_3^{(\mathrm{S,inh})}}{dz} + i\Delta k_z + iz\frac{d\Delta k_z}{dz}\right) \qquad (18b)$$

where the superscript (S,inh) represents the dependence of the functions on the saturation parameter and the inhomogeneous character of the spectral linewidth selected in this model study, respectively. In general, the functionality of the coefficients is

$$\Psi_1^{(\mathrm{S,inh})}(z) = f\left(S(z), \frac{dS}{dz}(z), z\right); \quad \Psi_2^{(\mathrm{S,inh})}(z) = q\left(S(z), \frac{dS}{dz}(z), z\right), \qquad (19)$$

where $\frac{dS}{dz} = -2\alpha_1 S$ is the expression for the saturation parameter. Considering the expression for the absorption coefficients, this derivative is given by

$$\frac{dS}{dz} = -\frac{\pi N \mu_{ab}^2 \omega_1 T_2 \rho_D^0}{\hbar c \eta_1} \times \left\{ \frac{\left[ \begin{array}{c} c_0 + c_1(4S+1)^{1/2} + c_2(4S+1) + c_3(4S+1)^{3/2} \\ + c_4(4S+1)^2 + (4S+1)^{5/2} \end{array} \right]}{\left[ d_0(4S+1)^{1/2} + d_1(4S+1)^{3/2} + (4S+1)^{5/2} \right]} \right\}, \tag{20}$$

where the coefficients are given by

$$\begin{aligned} c_0 &= -\gamma T_2^3 \left[ \gamma^2 + (\omega_c - \omega_1)^2 \right]; \quad c_1 = T_2^2 \left[ \gamma^2 - (\omega_c - \omega_1)^2 \right] \\ c_2 &= \gamma T_2 \left\{ T_2^2 \left[ \gamma^2 + (\omega_c - \omega_1)^2 \right] + 1 \right\}; \\ c_3 &= -\left\{ T_2^2 \left[ \gamma^2 - (\omega_c - \omega_1)^2 \right] + 1 \right\}; \quad c_4 = -\gamma T_2; \\ d_0 &= T_2^4 \left[ \gamma^2 + (\omega_c - \omega_1)^2 \right]^2; \quad d_1 = -2 T_2^2 \left[ \gamma^2 - (\omega_c - \omega_1)^2 \right]. \end{aligned} \tag{21}$$

In this work, we have solved (17) numerically for the amplitude $\vec{E}_3(z)$ in order to calculate the quotient between the intensities of the signal and the probe at the beginning of the optical length, according to $I_3(z)/I_2(0) = |E_3(z)|^2/|E_2(0)|^2 = X(\Delta_1, \Delta_2; \Re)$. We represent the solution of $X$ as a function of $\Delta_1$ and $\Delta_2$ in the frequency space, parametrized by $\Re$ (where $\Re = N, S, z, \gamma, T_1, T_2$).

### 2.2.2. *Topological studies for the FWM signal surfaces*

Figure 2 shows four surfaces for the intensity of FWM as a function of $\gamma$ and $z$. At any given inhomogeneity, the signal intensity grows from the origin of the cell until it reaches a maximum and then gradually decreases. The distribution around the maximum intensity is broadened as $\gamma$ increases, which is more noticeable

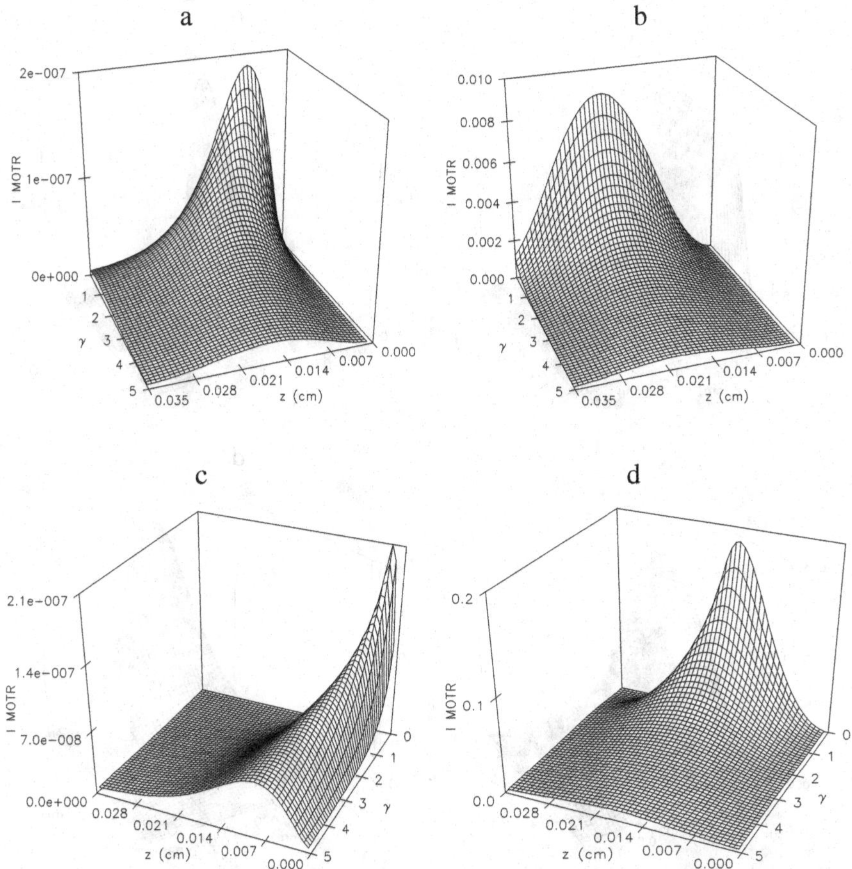


**Fig. 2.** Intensity surfaces for the FWM signal. Cases (a) $S(0) = 0.0006$, $N = 0.25 \times 10^{-3}$ M; (b) $S(0) = 2.0$, $N = 0.25 \times 10^{-3}$ M; (c) $S(0) = 0.0006$, $N = 1.00 \times 10^{-3}$ M; (d) $S(0) = 2.0$, $N = 1.00 \times 10^{-3}$ M.

at larger concentration values (Figs. 2(c) and 2(d)). It is also important to notice the decrease in the intensity of the signal as $\gamma$ increases under conditions of low concentrations (Figs. 2(a) and 2(b)).

Figure 3 depicts four intensity surfaces for the FWM signal as a function of the concentration $N$ and optical path ($z$). A fixed value of $\gamma = 3.12(1/T_2)$ was used in Figs. 3(c) and 3(d), which is the value reported for Malachite green in water. The intensity

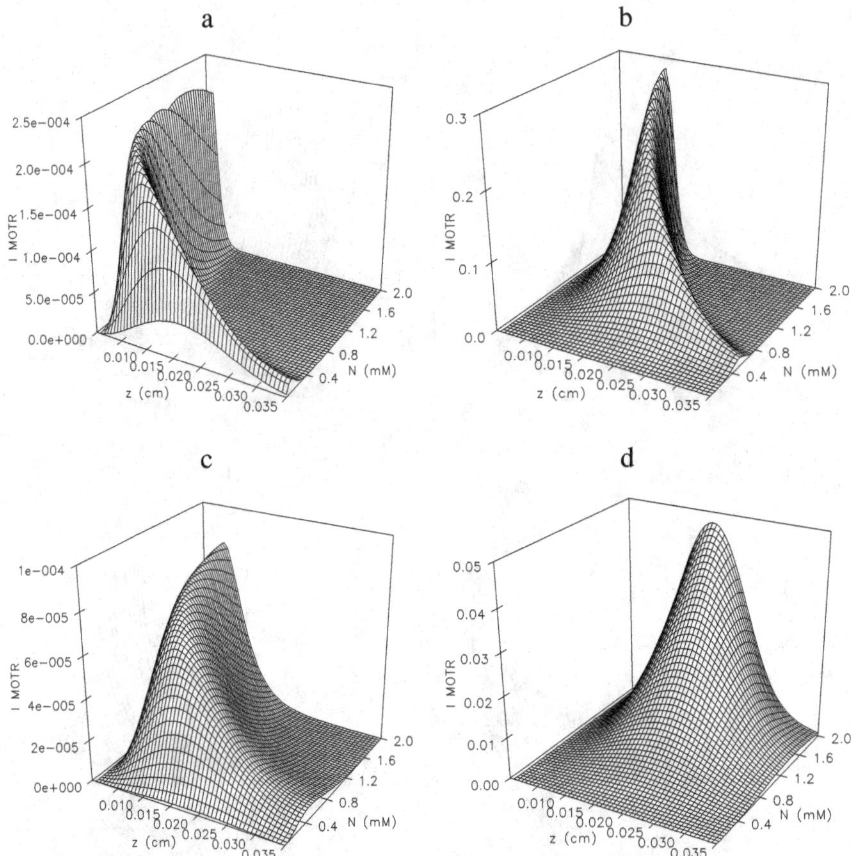


**Fig. 3.** Intensity surfaces for the FWM signal. Cases: (a, c) $S(0) = 0.02$; (b, d) $S(0) = 2.0$; (a, b) $\gamma = 0.0001(T_2^{-1})$; (c, d) $\gamma = 3.12(T_2^{-1})$.

of the signal reaches a maximum along the $z$ for any value of $N$, and that the maximum values are closer to the origin of the cell as the concentration increases. This effect is more noticeable at low values of $\gamma$ (Figs. 3(a) and 3(b)) and at lower pump intensities (Figs. 3(a) and 3(c)). Figures 3(a)–3(c) show that for fixed values of the length of the cell, and at values far from the origin, the intensity of the signal reaches a maximum at larger values of $N$ as the optical path becomes shorter, which is indicative of an absorption process.

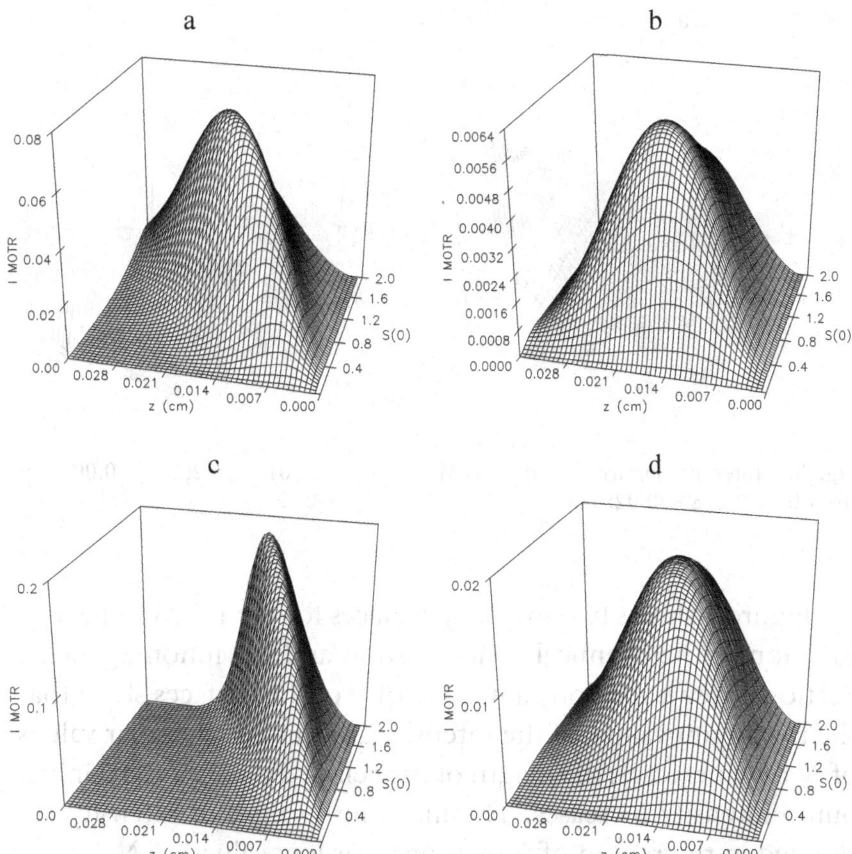


**Fig. 4.** Intensity surfaces for the FWM signal. Cases: (a, b) $N = 0.5\,\text{mM}$; (c, d) $N = 1.0\,\text{mM}$; (a, c) $\gamma = 0.0001(T_2^{-1})$; (b, d) $\gamma = 3.12(T_2^{-1})$.

Figure 4 represents four intensity for the FWM signal as a function of the initial saturation parameter $S(0)$ and the size of the cell $z$. This intensity behavior is larger at low values of $\gamma$ (Figs. 4(a) and 4(c)). However, further increases in the pump intensity, on saturation conditions, do not appreciably affect the position of the intensity maximum. At this point it is worth mentioning that an increase in the $N$ originates equivalent effects as decreasing $\gamma$, although both properties have a different nature.

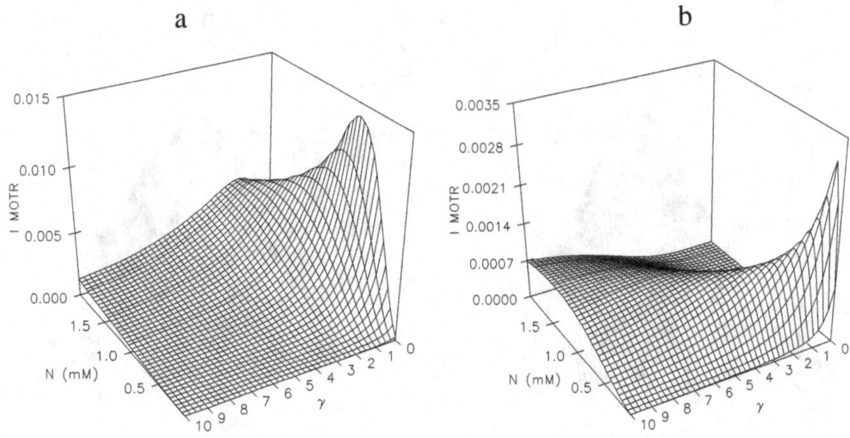


**Fig. 5.** Intensity surfaces for the FWM signal. (a) $S(0) = 0.2$, $z = 0.005$ cm; (b) $S(0) = 0.2$, $z = 0.025$ cm.

Figure 5 shows two intensity surfaces for the FWM signal as a function of the chemical concentration and the inhomogeneity. Sectional views at a constant value of $\gamma$ of the surfaces show that the maximum values of the intensity are shifted to higher values of $N$ as $\gamma$ increases, the length of the cell decreases, and the initial pump intensity increases. The maximum values of intensity are reached in the region of $\gamma \to 0$ only for low values of $N$.

Figure 6 depicts four surfaces showing the dependency of the intensity surface for the FWM signal with the initial saturation parameter and the spectral inhomogeneity broadening. Here, cuts at $\gamma$ constant of these surfaces show that the maximum values of the intensity are shifted to higher values of pump beam intensity as $\gamma$ decreases. The maximum intensity of these surfaces is displaced to low inhomogeneity values by decreasing the size of the cell (Figs. 6(a) and 6(c)) or by decreasing the solute concentration (Figs. 8(a) and 8(b)). It could also be observed that saturation of the signal occurs at lower initial pump intensities when both $N$ and the optical path decrease.

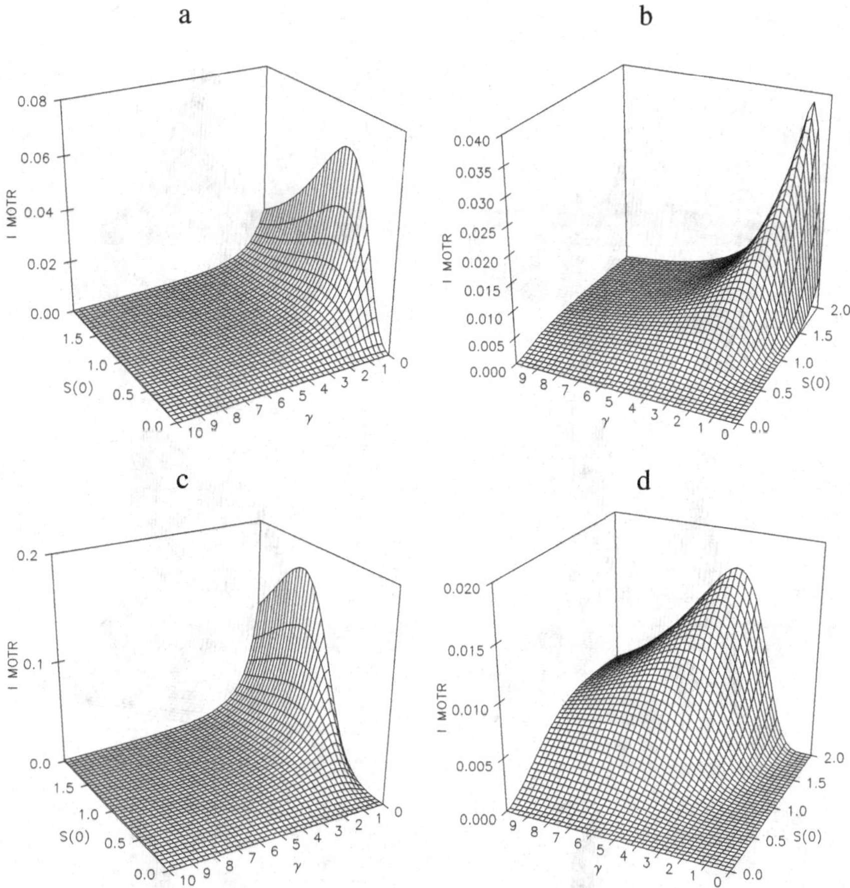


**Fig. 6.** Intensity surface for the FWM signal. Cases: (a, b) $N = 1.0\,\text{mM}$; (c, d) $N = 2.0\,\text{mM}$; (a, c) $z = 0.005\,\text{cm}$; (b, d) $z = 0.025\,\text{cm}$.

### 2.2.3. *Spectra in the frequency space*

The frequency spectra shown in Figs. 7 and 8 are obtained in the present work when the initial pump's intensity is low enough to treat the pump's field through second order in perturbation. These spectra show the formation of an absorption hole about the resonant frequency of the pump field when the cell's size is increased. It is also noticeable from Fig. 7 that the spectral symmetry is broken as the optical length of the beams is increased.

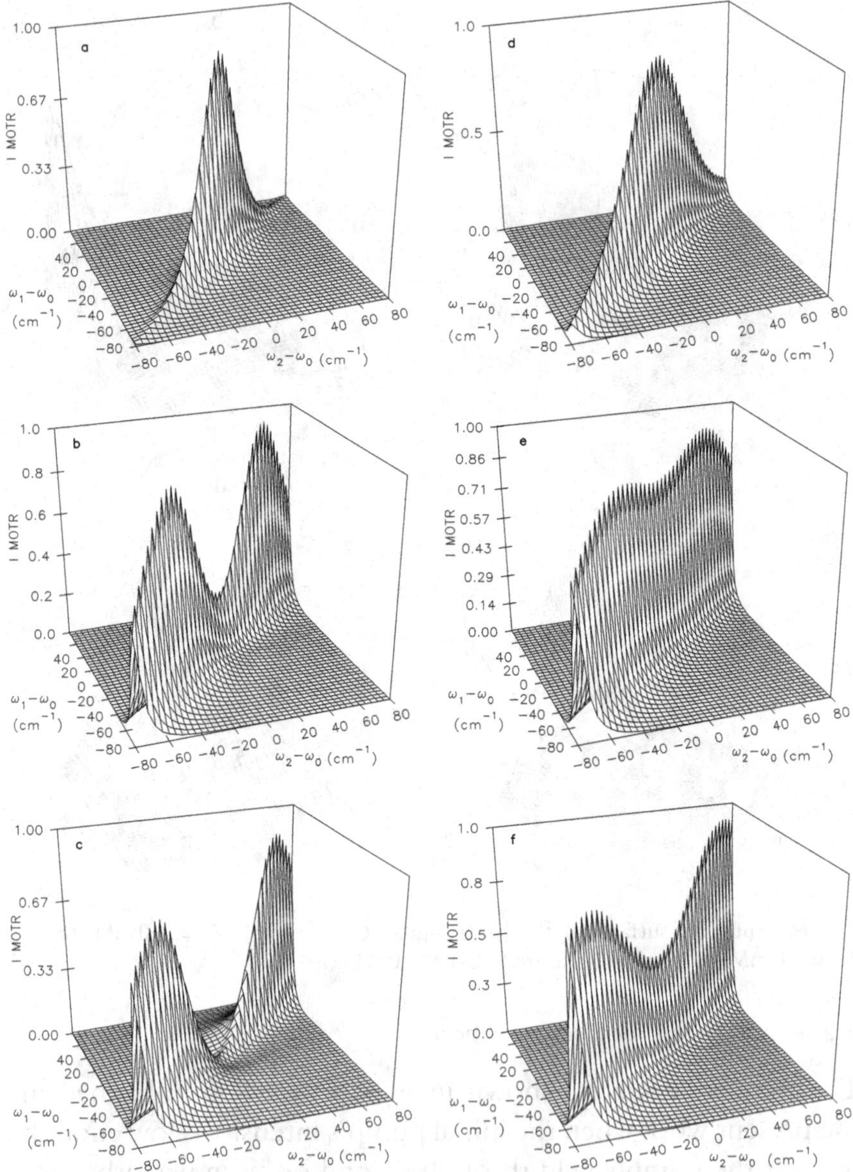

**Fig. 7.** Intensity of the FWM signal in the frequency space. $T_1 = 10T_2$, $S(0) = 0.02$, $N = 0.5$ mM. Cases: (a, b, c) $\gamma = 0.0001(T_2^{-1})$; (d, e, f) $\gamma = 0.75(T_2^{-1})$; (a, d) Local case; (b, e) $z = 0.015$ cm; (c, f) $z = 0.025$ cm.

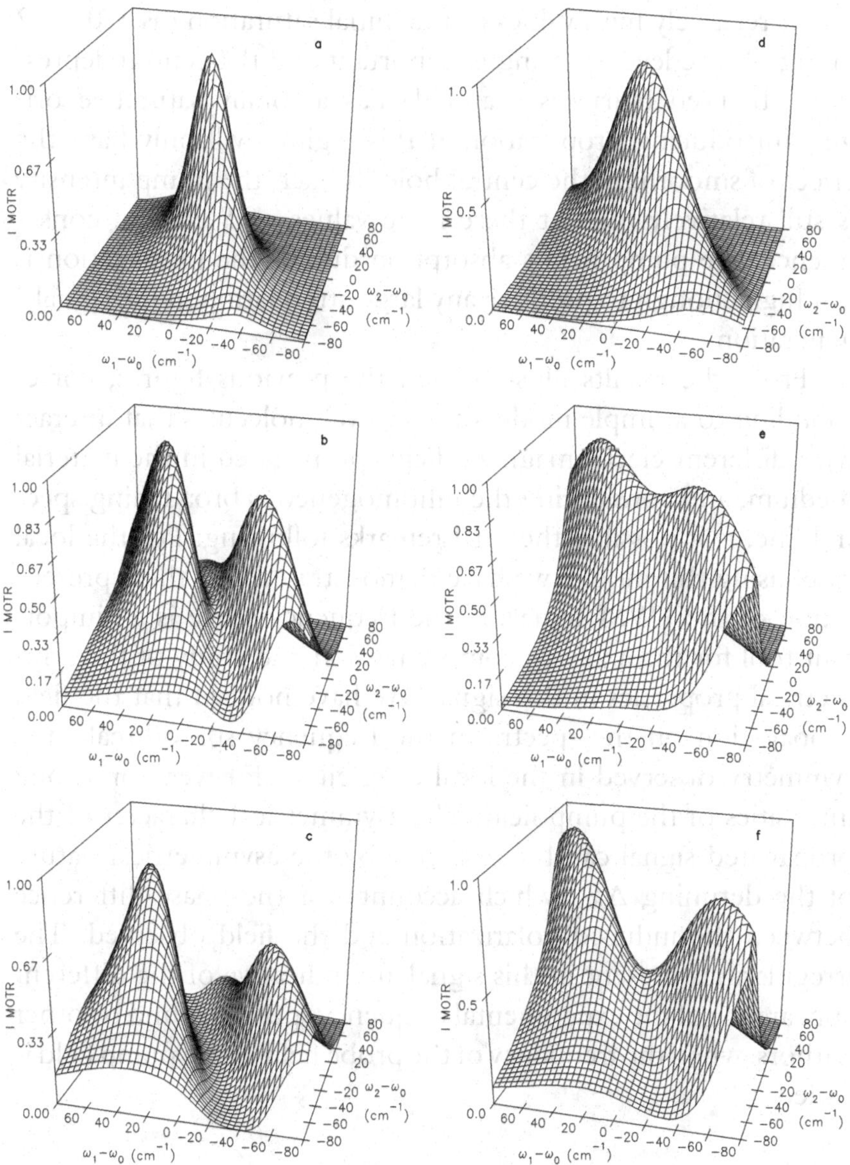


**Fig. 8.** Intensity of the FWM in the frequency space. $T_1 = T_2$, $S(0) = 0.02$, $N =$ 0.5 mM. (a, b, c) $\gamma = 0.0001(T_2^{-1})$; (d, e, f) $\gamma = 0.75(T_2^{-1})$; (a, d) local case; (b, e) $z = 0.015$ cm; (c, f) $z = 0.025$ cm.

At relatively high values of the initial saturation (as $S(0) = 2$ in Fig. 9) the level splitting is important and the central depression observed (surfaces a and d) has a totally saturative origin. Introducing propagation at this regime will only have the effect of smoothing the central hole. In fact, the pump intensity is still relatively high at these large values of $S$ and, in consequence, the pump field's absorption due to the propagation is not high enough to produce any large attenuation in the signal's generation.

From the results illustrated in the previous figures, corresponding to a simple model of two-level molecules that interact with different electromagnetic fields propagated in the material medium, and considering the inhomogeneous broadening spectral line, is it possible the final remarks following: For the local case, using this model, we have demonstrated that the representation of the FWM signals in the frequency space is an important tool for the analysis, comprehension and study of the symmetrical properties in this signal. We have noticed that the field propagation on the spectra in the frequency space breaks the symmetry observed in the local case, either for weak or strong intensities of the pump field. The asymmetrical character of the propagated signal can be explained by the asymmetrical nature of the detuning $\Delta k_z$, which accounts for the phase difference between the induced polarization and the field observed. The irregularities present in this signal are indicative of the different appearance of the experimental frequency spectra, found by other authors, when the frequency of the probe field or the pump field is varied.

### 2.3. *Approximation levels for the study of the propagation in FWM*

In a propagation treatment at all orders in the pump field and to first order in the probe and signal fields, we obtain three

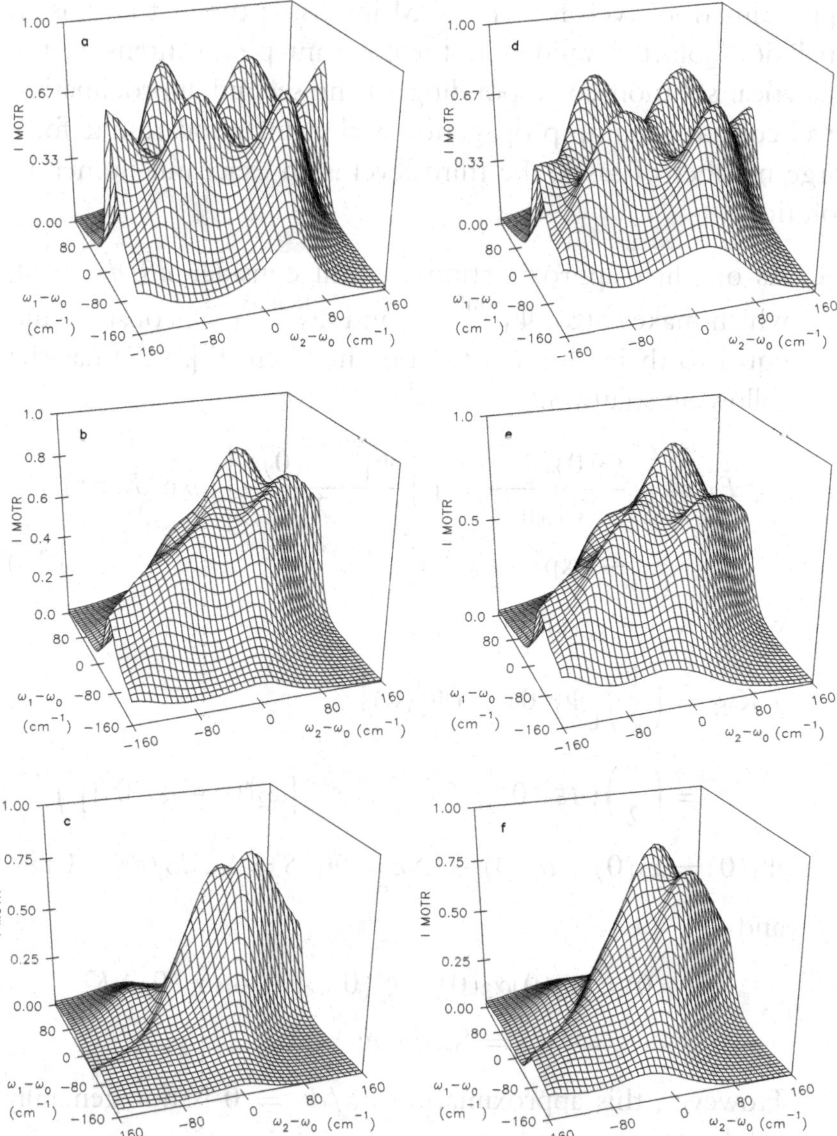


**Fig. 9.** Intensity of the FWM in the frequency space. $T_1 = T_2$, $S(0) = 2.0$, $N = 0.5\,\text{mM}$. (a, b, c) $\gamma = 0.0001(T_2^{-1})$; (d, e, f) $\gamma = 0.75(T_2^{-1})$; (a, d) local case; (b, e) $z = 0.015$ cm; (c, f) $z = 0.025$ cm.

approximation levels for the FWM intensity: the first level is an analytical solution valid only for constant pump intensity; the analytical solution corresponding to the second approximation level considers pump propagation and yields good results for a large number of cases; the third level is an exact but numerical solution.

(A) As our first approximation we will consider $dS/dz = 0$, which makes $S(z)$, $\Psi_1^{(S,\mathrm{inh})}(z)$ and $\Psi_2^{(S,\mathrm{inh})}(Z)$, constant and equal to their values at the origin. Then, Eq. (17) has the following solution:[7]

$$\tilde{E}_3(z) = \frac{\xi_3(0)\tilde{E}_2^*(0)}{2K_{\mathrm{eff}}} \exp\left[\frac{\Psi_1^{(S,\mathrm{inh})}(0)z}{2}\right] \left[\exp\left(K_{\mathrm{eff}} z\right) - \exp\left(-K_{\mathrm{eff}} z\right)\right], \tag{22}$$

where $K_{\mathrm{eff}}$ is given by

$$K_{\mathrm{eff}} = \left(\frac{1}{2}\right)\left[\Psi_1^2(0) - 4\Psi_2^2(0)\right]^{1/2} = \left(\frac{1}{2}\right)\left(4\xi_2^*(0)\xi_3(0) + \left[i\Delta k - \left[\alpha_2(0) - \alpha_3(0)\right]\right]^2\right)^{1/2},$$

$$\Psi_1(0) = \alpha_2(0) + \alpha_3(0) - i\Delta k = \Psi_1(S = S_0, dS/dz = 0)$$

and

$$\Psi_2(0) = \alpha_2(0)\alpha_3(0) - \xi_2^*(0)\xi_3(0) - i\alpha_3(0)\Delta K = \Psi_2(S = S_0, dS/dz = 0).$$

However, this approximation $dS/dz = 0$ is not generally valid.

(B) As our second approximation we will consider $dS/dz = -2\alpha_1(z)S(z)$, where $S(z)$ is given by $S(z) = (\vec{\mu}_{ba} \cdot \vec{E}_1/\hbar)^2 T_1 T_2$. However, it may just be that both $\Psi_1(z)$ and $\Psi_2(z)$ are still practically constant along the optical path despite the fact that $dS/dz \neq 0$ because either the optical length is small enough or the $z$ dependence of the functions $\Psi_1(z)$ and $\Psi_2(z)$ is rather weak. In this case $\Psi_1(z)$ and $\Psi_2(z)$ are again constant and equal to their values at the origin, and Eq. (24) is also valid in this approximation level, but now with:

$$\tilde{E}_3(z) = \left[\frac{\alpha_3(0)\vec{E}_3(0) - \xi_3(0)\vec{E}_2^*(0) + E_3(0)\beta_2}{\beta_2 - \beta_1}\right]\exp(\beta_1 z) + \left[\frac{-\alpha_3(0)\vec{E}_3(0) + \xi_3(0)\vec{E}_2^*(0) - E_3(0)\beta_1}{\beta_2 - \beta_1}\right]\exp(\beta_2 z), \tag{23}$$

where $\beta_2 = -\frac{\Psi_1(0)}{2} - K_{\text{eff}}$; $\beta_1 = -\frac{\Psi_1(0)}{2} + K_{\text{eff}}$ ($z = 0$ refers in this case to the origin of each microcell, is then introduced as the input of the next microcell).

To determine how many microcells are necessary for a given set of experimental conditions, we divide the optical length first into two microcells and then into larger numbers of cells, until the result of the calculation does not show a significant change as the number of microcells is increased. This last calculation is the correct one and yields the minimum number of microcells that need to be used.

(C) The third level of approximation, correspond to a division of the optical length into enough microcells to make the second level of approximation valid inside each. Then Eqs. (22) and (23) are used repeatedly with the appropriate boundary conditions.

## 3. Intramolecular Coupling

### 3.1. *Molecular models*

In general, the two-level model description of atoms and molecules interacting with a classic electromagnetic field, where the interaction is represented by the coupling of the field to the transition dipole moments of the system, have shown great utility in nonlinear optics. In this respect, there are many studies in the literature where a two-levels model (without internal structure) is employed in diatomic molecules and dimers. However, such states in a polyatomic molecule can be thought as vibrational states belonging to different potential energy surfaces. These models involving two degenerate or quasi-degenerate electronic curves in a small range of nuclear coordinates have been applied to Jahn–Teller and pseudo-Jahn–Teller coupling in molecules and to vibronic coupling on degenerated excited states of dimers.[36]

Theoretical descriptions of photophysical and photochemical processes in the literature involve the treatment of two or more electronic states and several or many vibrational modes (for large molecules) with strong electronic–vibrational interactions.[37] The "vibronic coupling," resulting from the coupling between the nuclear and electronic motions in a molecule is of great importance to some physical chemistry processes, for instance, in the study of relaxation rates of internal conversion,[28] in the study of weak transition probabilities forbidden by symmetry in absorption and emission processes,[27] in femtosecond spectroscopy,[29] in the study of the optical absorption band shape of dimers[30] and in the study of resonances in scattering processes.[31]

The principal aim of the present contribution is to show the effects of changes in the coupling parameters $v$ and $V_0$, on the dipole moments of the coupled states, and in consequence, on the maximal intensity values of the Four-Wave Mixing (FWM) response, taking into account different values of the strength constant $\delta$.

The present model is described as an ensemble of two-levels systems conformed by ground vibrational energy states belonging to two crossed harmonic potential energy curves with different apertures, as measured by the strength constant ($\delta$), which have their minima horizontally and vertically displaced in their nuclear coordinate ($R$) and energy ($V_0$), respectively. The molecular systems interact with a classic electromagnetic fields treated as plane wave, where we have included a relaxation mechanism associated to the presence of the solvent, which is treated as transparent to the radiation and it is introduced in a phenomenological way. By using the dipole radiation–matter interaction in our model, the permanent dipole moments of states in the uncoupled basis are also included. It has been demonstrated that they contribute significantly to the photonic processes that take place outside the resonant region of the spectrum.[10,11,22,25] The general effects on changing the coupling parameter on the global FWM signal spectra were shown in the previous work.[38,39] They indicated that the critical quantities on the study of the FWM response in a coupled basis were the transition and the permanent dipole moments when the rotating wave approximation (RWA) was not taken into account. This approximation was not introduced at present in order to study physical processes that occur out of the resonance region.

### 3.2. *Theoretical characteristics of the model*

The two-levels system employed in this chapter is described using a coupled-basis model including two crossed harmonic potential energy curves, which are displaced horizontally in nuclear coordinate and vertically in energy, where the spin–orbit interaction is treated as a perturbation in the global Hamiltonian describing the molecular systems. Each potential energy curve can have different apertures and they include only their fundamental vibrational levels. As indicated in Fig. 10, the following parameters are considered: the coupling parameter ($v$), the energy difference between the minima of the potential-energy curves ($V_0$), the energy height at which the coupling occurs ($S$), and the relative aperture of the potential energy curves involved ($\delta$).

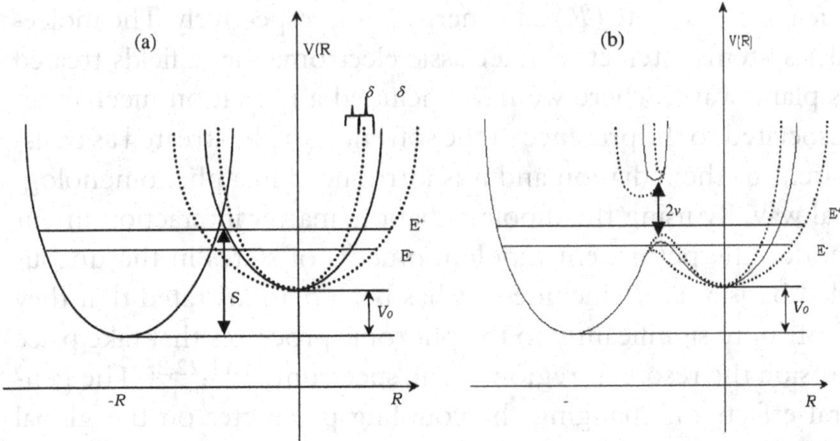


**Fig. 10.** (a) Diabatic and (b) adiabatic representation of the two crossed harmonic potential curves, representing the intramolecular coupling for different apertures (dot lines).

Each electronic and vibrational state is described by its respective wave function and energy value. Taking a linear combination of the eigenfunctions of each level and solving for the secular determinant, it is always possible to generate the eigenfunctions and eigenvalues of the coupled states (considering $-(+)$ the low (high) new coupled states):

$$E^{\pm} = \frac{1}{2}\left[(E_{10} + E_{20}) \pm \left[\Delta E^2 + 4\,|\,V_{00}|^2\right]^{1/2}\right], \quad (24)$$

$$\begin{aligned}\Psi^{\pm}(r;R) = \frac{1}{C_{10}^{\pm}}\,&[|\,V_{00}|\psi_1(r;R)\Phi_{10}(R)\\ &\pm (E_{10} - E^{\pm})\psi_2(r;R)\phi_{2k}(R)], \end{aligned} \quad (25)$$

where the vibrational energies are $E_{10} = 0.5$ and $E_{20} = 0.5\delta + V_0$, $\Delta E = (E_{20} - E_{10})$, $V_{00} = v\langle\varphi_{10}|\varphi_{20}\rangle$, $C_{10}^{\pm} = [|\,V_{00}|^2 + (E_{10} - E^{\pm})^2]^{1/2}$, the overlap integral, obtained by the Pekarian formula,[40] is given by $\langle\varphi_{10}|\varphi_{20}\rangle = \frac{(4\delta)^{1/4}}{(1+\delta)^{1/2}} \exp\{-\frac{S}{2}[1 - (1+\delta)^{-1}]\}$, $\delta = \tilde{\omega}_0/\omega_0$ and $S = \left(\frac{\tilde{m}\omega_0}{\hbar}\right) R_0^2$. Here, $\tilde{m}$ represents the reduced mass associated with the vibrational modes as described by the molecular coordinate $R_0$ and frequency $\omega_0$.

In the new basis of coupled states, there exist a new set of dipole moments that are different to those in the uncoupled basis. To know the new expressions to the new coupled basis, the following integral should be calculated:

$$m_{ij}(R) = \int \Psi_i^*(r;R)\hat{M}\Psi_j^*(r;R)d^3r, \tag{26}$$

where $\hat{M}$ is the total electronic dipole moment operator.

It has been shown that the zero values of the dipole moments in the uncoupled basis do not imply the nullity of the dipole moments in the new generated basis.[37] They are in fact described by the following expressions:

$$\mu_{-+} = \left\{\frac{|V_{00}|^2}{4\,|V_{00}|^2 + \Delta E^2}\right\}^{1/2}\left[(m_{11} - m_{22}) + m_{12}\frac{\Delta E}{v}\right], \tag{27}$$

$$\mu_{aa} = \left\{\frac{|V_{00}|^2}{2\,|V_{00}|^2 - \Delta E\,(E_{10} - E^a)}\right\}\left\{m_{22} + m_{11} - \frac{2\,(E_{10} - E^a)\,m_{12}}{v}\right\}$$
$$- \frac{(E_{10} - E^a)\,\Delta E m_{22}}{2\,|V_{00}|^2 - \Delta E\,(E_{10} - E^a)}, \tag{28}$$

where $m_{11}$, $m_{22}$, and $m_{12}$ represent the permanent and transition dipole moments of states in the uncoupled basis, respectively. As we shall show below, these quantities are very important to determine the behavior of the nonlinear signal because of their dependence to the Macroscopic Polarization, and consequently with the intensity of the signal studied.

### 3.3. *Signal response*

The Liouville formalism is the most common mathematical method employed to study the interaction of the states in the new basis with the electromagnetic field using the macroscopic polarization. The equation $i\hbar\frac{d\rho(t)}{dt} = [H, \rho(t)]$ represents the start point to obtain the Optical Bloch Equations (OBE), which describe the temporal evolution of the system. In our case, the OBE are obtained by considering the following aspects:

(*a*) Use of a solvent transparent to the radiation, included in the calculation in a phenomenological way; (*b*) consideration of dipole field–matter interaction; (*c*) inclusion of the secular approximation; (*d*) neglecting the Rotating Wave Approximation (RWA), which allows us to study the signal response out of the resonance frequency; (*e*) explicit inclusion of the permanent dipole moments of states in the uncoupled basis.

Under conditions mentioned above, the OBE assume the following form:

$$\frac{d\rho_{-+}}{dt} = -\frac{i}{\hbar}H_{-+}\rho_D - \frac{i}{\hbar}\rho_{-+}\left[H_{--} - H_{++}\right] - \left(\frac{1}{T_2} + i\omega_0\right)\rho_{-+}, \tag{29}$$

$$\frac{d\rho_{+-}}{dt} = \frac{i}{\hbar}H_{+-}\rho_D + \frac{i}{\hbar}\rho_{+-}\left[H_{--} - H_{++}\right] - \left(\frac{1}{T_2} - i\omega_0\right)\rho_{+-}, \tag{30}$$

$$\frac{d\rho_D}{dt} = -\frac{2i}{\hbar}\left(H_{+-}\rho_{-+} - \rho_{+-}H_{-+}\right) - \frac{1}{T_1}\left[\rho_D - \rho_D^0\right], \tag{31}$$

where $\rho_{-+}$, $\rho_{+-}$ are the coherences elements (non-diagonal) of density matrix and $\rho_D$ is the difference between the populations elements (diagonal) of density matrix, respectively.

The above differential equations represent the temporal evolution of the coherence (Eqs. (29) and (30)) and populations (Eq. (31)), and they include a term related to the molecular system (associated with the resonance frequency $\omega_0$), a term related to the field–system interaction (described by the dipole

Hamiltonian) and a term related to the relaxation process, characterized by the presence of the longitudinal and transversal relaxation times $T_1$ and $T_2$, respectively.

Considering the OBE established before for this particular case (Eqs. (29)–(31)), it is possible to obtain an expression for the magnitude of the polarization, and consequently, for the signal intensity by using the relationship $I = \frac{c\varepsilon}{8\pi}|P|^2$. This expression can be written as

$$P(\omega_3) = N\langle\mu\rangle = N\begin{pmatrix}\mu_{--} & \mu_{-+}\\ \mu_{+-} & \mu_{++}\end{pmatrix}\begin{pmatrix}\rho_{--} & \rho_{-+}\\ \rho_{+-} & \rho_{++}\end{pmatrix}$$
$$= N\left[\mu_{-+}\rho_{+-}(\omega_3) + \mu_{+-}\rho_{-+}(\omega_3) - d_{ic}\rho_D(\omega_3)\right]. \quad (32)$$

Here, $N$ represents the chemical concentration of the absorbent molecules, $\mu$ represents the dipole moments, and $\rho$ represents the density matrix elements, all of them corresponding to the coupled basis. Solving the OBE perturbatively (to third order in the total field), we can obtain our final expression for the Macroscopic Polarization to the signal frequency $\omega_3$:

$$P(\omega_3) = Ni\rho_D^{(0)}E_1^2E_2^*\left\{2\mu_{-+}^4\left(\frac{1}{D_3^-} - \frac{1}{(D_3^+)^*}\right)\left[\frac{1}{\Gamma}\Lambda_1 + \frac{1}{\lambda}\Lambda_2\right]\right.$$
$$\left. + d_{ic}^2\mu_{-+}^2\left[\left(\frac{1}{\beta} - \frac{1}{(D_3^+)^*}\right)\Lambda_3 + \left(\frac{1}{D_3^-} - \frac{1}{\beta}\right)\Lambda_4\right]\right\}, \quad (33)$$

where we have defined the following terms:

$$\Lambda_1 = \left[\frac{1}{D_2^+} + \frac{1}{D_1^-} + \frac{1}{(D_2^-)^*} + \frac{1}{(D_1^+)^*}\right],$$

$$\Lambda_2 = \left[\frac{1}{D_1^-} + \frac{1}{(D_1^+)^*}\right],$$

$$\Lambda_3 = \left[ \frac{1}{(D_\Delta^+)^* (D_2^-)^*} + \frac{1}{(D_1^+)^* (D_\Delta^+)^*} + \frac{1}{(D_1^+)^* (D_5^+)^*} \right],$$

$$\Lambda_4 = \left[ \frac{1}{D_\Delta^- D_2^+} + \frac{1}{D_1^- D_\Delta^-} + \frac{1}{D_1^- D_5^-} \right],$$

$$\vec{d} = \vec{\mu}_{--} - \vec{\mu}_{++}, \quad \rho_D = \rho_{++} - \rho_{--}, \quad \Delta = \omega_1 - \omega_2,$$

$$\Gamma = \frac{i}{T_1} - i(\omega_1 - \omega_2), \quad \beta = \frac{1}{T_1} - i\omega_3, \quad \lambda = \frac{1}{T_1} - 2i\omega_1,$$

$$D_j^\pm = \frac{1}{T_2} + i(\omega_0 \pm \omega_j), \quad D_5^\pm = \frac{1}{T_2} + i(\omega_0 \pm 2\omega_1)$$

and where $T_1$ and $T_2$ represent the longitudinal and transversal relaxation times, respectively.

### 3.4. *Results and discussion*

In this section we will show some results for the FWM signal response obtained when changes in the coupling parameters $\nu$ and $V_0$ occur and taking into account different values of $\delta$. The intramolecular coupling parameters employed in this work were varied in the range of $\nu$ from 0.01 to 0.5 and $V_0$ from 0.01 to 1. Other important parameters to be considered are: ($a$) $\omega_0 = 3.0628 \times 10^{15}\,\text{s}^{-1} = 16{,}280\,\text{cm}^{-1}$ (resonance frequency of an organic molecule — Malachite Green), ($b$) the longitudinal and transversal relaxation times $T_1 = T_2 = 1.3\times 10^{-13}$ s, ($c$) $S = 0.1$, ($d$) $m_{11} = 1$ D and $m_{22} = 1.3$ D, respectively, for the permanent dipole moments of the uncoupled states (giving as a result $d = 0.3$ D for the difference in permanent dipolar moments) and ($e$) $m_{12} = m_{21} = 0.1$ D for the transition dipole moments of the uncoupled states.

Figures 11 and 12 depict the three-dimensional representation of the transition and permanent dipole moments of the coupled states, respectively, as a function of the coupling parameters

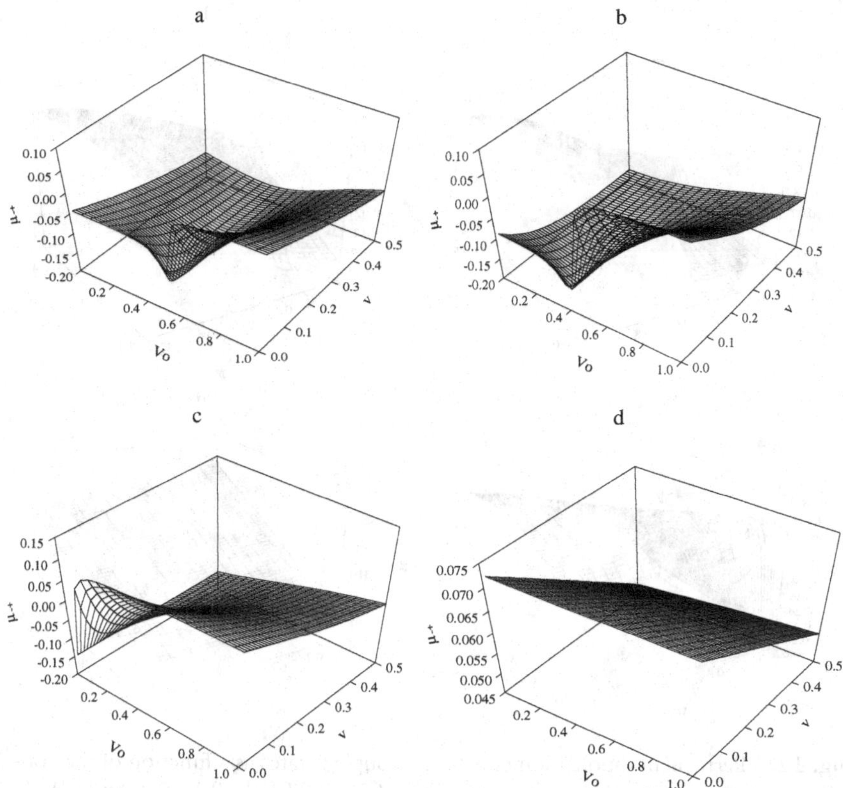


**Fig. 11.** Transition dipole moments of the coupled states as a function of the coupling parameters $\nu$ and $V_0$ for different values of $\delta$ (a) 0.01; (b) 0.1; (c) 1, and (d) 10.

$\nu$ and $V_0$, at values of $\delta$ changing in different orders of magnitude: (*a*) $\delta = 0.01$, (*b*) $\delta = 0.1$, (*c*) $\delta = 1$, and (*d*) $\delta = 10$. There are two important facts to be mentioned related to these graphs. First, we can observe that the transition and permanent dipole moments vanish for particular values of $\nu$, $V_0$, and $\delta$. In second place, we find a dramatic change in the behavior of the dipole moments at high values of $\delta$, where they can take constant values for different values of $V_0$ and $\nu$.

Based on this behavior, it is possible to find mathematical expressions that permit us to establish the values of the parameters $\delta$, $\nu$, and $V_0$ for which the dipole moment take a null value.

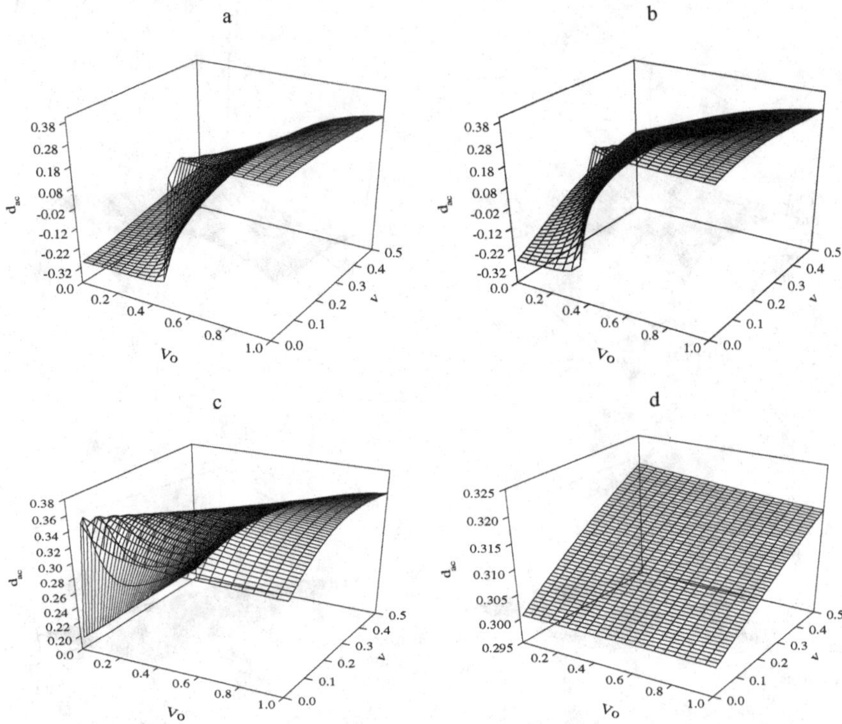


**Fig. 12.** Permanent dipole moments of the coupled states as a function of the coupling parameters $\nu$ and $V_0$ for different values of $\delta$: (a) 0.01 (b) 0.1 (c) 1, and (d) 10.

Inspecting Eqs. (27) and (28) it is possible to find that for the transition dipole moment, the null value is related with the nullity of the term. Manipulating this difference and substituting for the values of the dipole moments in the uncoupled basis, we find that the nullity of the transition dipole moment is obtained when:

$$\delta = 1 + 6\nu - 2V_0. \tag{34}$$

We can proceed in a similar way for the permanent dipole moments. The nullity of Eq. (28), occurs in two cases: (*a*) when $\mu_{--} = \mu_{++}$, and (*b*) when both are zero ($\mu_{--} = \mu_{++} = 0$). Due to the difficulty to find a clear expression for first case, we have payed attention to the second case, i.e, $\mu_{--} = \mu_{++} = 0$, which occures when:

$$\left\{\frac{1}{|V_{00}|^2}\left[-\left(E_{10}-E^a\right)\Delta E+|V_{00}|^2\right] m_{22} + m_{11} - \frac{2\left(E_{10}-E^a\right) m_{12}}{\nu}\right\} = 0. \tag{35}$$

Manipulating this expression after substituting the values of the dipole moments in the uncoupled basis, it is possible to obtain:

$$\left[\frac{1}{8}(\delta-1)^2+\frac{1}{2}V_0(V_0+\delta-1) + |V_{00}|^2\left(2.3\nu-0.1\left(\frac{1}{2}-\frac{1}{2}\delta-V_0\right)\right)\right] = \pm\left[\frac{1}{4}-\frac{1}{4}\delta-\frac{1}{2}V_0-0.1\,|V_{00}|^2\right]\sqrt{\left(\frac{1}{2}-\frac{1}{2}\delta-V_0\right)^2+4\,|V_{00}|^2}. \tag{36}$$

It is important to notice that these expressions depend on the parameters $\nu$, $V_0$, and $\delta$. For each value of $\delta$, there exist values of $\nu$ and $V_0$ that induce the nullity of the transition and permanent dipole moment of states in the coupled basis. In general, changes generated in the transition and permanent dipole moments of states in the coupled basis due to the variations in the parameters $\nu$ and $V_0$ will induce modifications in the FWM signal response. In previous works, it was demonstrated that changes in the coupling parameters induce changes in the intensity and position of the resonances presented in the global FWM spectra.[38,39] At present, it is also possible to perform the same detailed study on each resonance in the global spectrum, which would allows us to generalize a common behavior of the FWM response. Modifications in the coupling parameters produce changes in the potential curves, and consequently, in its crossing. The resonance frequency of the new states changes consequently and also the nonlinear response of the system. These changes are associated with the behavior of the dipole moments of the states in the new basis, which is influenced by the modification in the charge distribution of the organic molecules with changes in the coupling parameters.

Finally, the intensity of the FWM signal predicted in the present model as a function of the pump and probe detuning is depicted in Fig. 13, where the intramolecular coupling is included in the frequency intervals $(-\omega_0, \omega_0)$ for $\omega_1$ and $(-2\omega_0, 2\omega_0)$ for $\omega_2$. It is important to note here that it is always possible to obtain a frequency spectrum with characteristics equivalent to

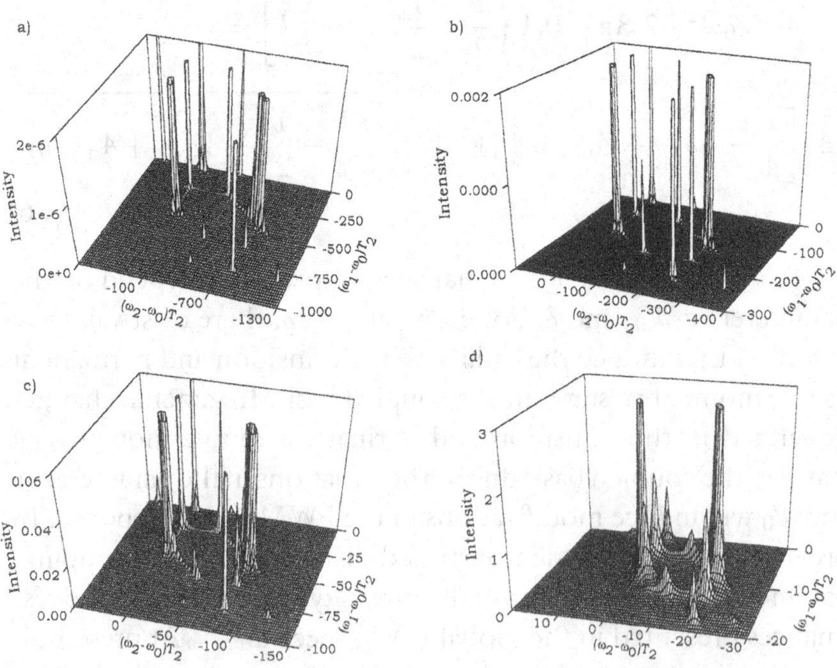


**Fig. 13.** Intensity of the FWM signal as a function of the pump $(\omega_1 - \omega_0)T_2$ and probe $(\omega_2 - \omega_0)T_2$ detuning, taking into account the intramolecular coupling for cases (a) $V_0 = 1$, (b) $V_0 = 0.3$, (c) $V_0 = 0.1$, and (d) $V_0 = 0.01$. The value of the intramolecular coupling parameter is 0.01 in all cases.

the previously studied cases using the present adiabatic representation by considering similar $\tilde{\omega}_0$ for the uncoupled basis.

We can observe from Fig. 13 that a shifting to lower detuning values of the different resonances involved with a decrease in the coupling parameters. For instance, the decrease in the parameter $V_0$ represented by the sequence from Figs. 13(a)–(d), generates a cluster formed by six out of a total of 12 peaks (Fig. 13(d)). As a highly evident from Fig. 13(d), the final spectrum has 12 resonance-structure configurations carrying the intensity symmetry and the symmetry of the detuning coordinates. Equivalent behavior can also occur when $V_0$ is constant and the parameter $\nu$ is decreased.

## 4. Final Remarks

The plot of the FWM signal intensity in the frequency space has proven to be useful in understanding the properties of spectra. This is due to the fact that the symmetrical characteristics of the nonlinear processes that give rise to the spectral line, lead to symmetrical properties in the frequency space, either for a second order regime or in all orders for the pump field. We have noticed that the field propagation on the spectra in the frequency space breaks the symmetry observed in the local case, either for weak or strong intensities of the pump field. The field propagation modifies the nature of the spectra in the frequency space due to absorption of the pump field. Here, the chemical concentration of the solution and the spectral inhomogeneity operate in opposite directions in the generation and characterization of any particular nonlinear signal.

For other part, the inclusion of intramolecular coupling on a two-levels model of a molecular system leads to changes in the intensity and position of the resonances presented in the global FWM spectra. Moreover, for specific values of $\nu$ and $V_0$, it is always possible to reproduce the same global spectrum found when the intramolecular coupling is not taken into account.[38,39]

In this work, it was possible to study the specific behavior of the maximum intensity of the FWM signal response under changes of the coupling parameters $\nu$, $V_0$, and $\delta$. These changes are related to modifications in the transition and permanent dipole moments of states in the new coupled basis.

Changes in the behavior of both transition and permanent dipole moments were obtained using different values of the coupling parameters. In this respect, we were able to find values of the dipole moments, which induce the nullity of the FWM signal at specific values of the vibronic coupling parameters. Mathematical relations are found to describe these results at specific values of the coupling parameters, which generate considerable changes in the FWM signal intensity.

## Acknowledgments

The present work was supported by the Fondo Nacional de Ciencia, Tecnología e Innovacion (FONACIT) (grant G-97000593) and by the Decanato de Investigaciones de la Universidad Simón Bolívar (grant GID-13).

## References

1. T. Yajima and H. Souma, *Phys. Rev. A* **17**, 309 (1978).
2. T. Yajima, H. Souma and Y. Ishida, *Phys. Rev. A* **17**, 324 (1978).
3. J. L. Paz, A. E. Cárdenas and A. J. Hernández, *Quantum Opt.* **5**, 355 (1993).
4. J. L. Paz, A. E. Cárdenas, A. J. Hernández and H. J. Franco, *Opt. Commun.* **109**, 195 (1994).
5. K. Wodkiewicz, *Phys. Rev. A* **19**, 1686 (1979).
6. G. Dattoli, S. Solimeno and A. Torre, *Phys. Rev. A* **35**, 1668 (1987).
7. R. Boyd, M. G. Raymer, P. Narum and D. Harter, *Phys. Rev. A* **24**, 411 (1981) and references therein.
8. I. Reif, F. García-Golding, J. L. Paz and H. J. Franco, *J. Opt. Soc. Am. B* **8**, 2470 (1991).
9. R. Bavli and Y. B. Band, *Phys. Rev. A* **43**, 5039 (1991).

10. J. P. Lavoine, C. Hoerner and A. A. Villaeys, *Phys. Rev. A* **44**, 5947 (1991).
11. R. Bavli and Y. B. Band, *Phys. Rev. A* **43**, 5044 (1991).
12. M. C. Bessega, J. L. Paz, A. J. Hernández and A. E. Cárdenas, *Phys. Lett. A* **206**, 305 (1995).
13. J. L. Paz, M. C. Bessega, A. E. Cárdenas and A. J. Hernández, *J. Phys. B.* **28**, 5377 (1995).
14. K. Shimoda and T. Shimizu, *Prog. Quant. Electron.* **2**, 1 (1972).
15. S. Leasure, K. Milfeld and R. E. Wyatt, *J. Chem. Phys.* **74**, 6197 (1981).
16. G. F. Thomas and W. J. Meath, *Phys. Lett. A* **70**, 396 (1979); *Mol. Phys.* **46**, 743 (1982); *J. Phys. B* **16**, 951 (1983).
17. B. Dick and G. Hohlneicher, *J. Chem. Phys.* **76**, 5755 (1982).
18. W. J. Meath and E. A. Power, *Mol. Phys.* **51**, 585 (1984); *J. Phys. B* **17**, 763 (1984).
19. M. A. Kmetic and W. J. Meath, *Phys. Lett. A* **108**, 340 (1985).
20. Y. B. Band, R. Bavli and D. F. Heller, *Chem. Phys. Lett.* **156**, 405 (1989).
21. R. Bavli and Y. B. Band, *Phys. Rev. A* **43**, 507 (1991).
22. R. Bavli, D. F. Heller and Y. B. Band, *Phys. Rev. A* **41**, 3960 (1990).
23. M. García-Sucre, J. L. Paz, E. Squitieri and V.Mujica, *Int. J. Quant. Chem.* **22**, 699 (1993).
24. M. García-Sucre, E. Squitieri, J. L. Paz and V. Mujica, *J. Phys. B* **27**, 4945 (1994).
25. M. A. Kmetic and W. J. Meath, *Phys. Rev. A* **41**, 1556 (1990).
26. M. Seel and W. Domcke, *J. Chem. Phys.* **95**, 7806 (1995) and references therein.
27. G. Herzberg and E. Teller, *Z. Phys. Chem. B* **21**, 410 (1933).
28. A. M. Mebel, M. Hayashi and S. H. Lin, *Chem. Phys. Lett.* **274**, 281 (1997).
29. W. Domcke and G. Stock, *Adv. Chem. Phys.* **100**, 1 (1997).
30. A. Pakhomov, S. Ekbundit, C. H. Lin, R. G. Alden and S. H. Lin, *J. Lummin.* **63**, 129 (1995).
31. D. M. Charutz, R. Baer and M. Baer, *Chem. Phys. Lett.* **265**, 629 (1997).
32. V. A. Yurovsky and A. Ben-Reuven, *J. Phys. B* **31**, 1 (1998).
33. M. D. levenson, *Introduction to Nonlinear Laser Spectroscopy* (Academic Press, New York, 1982).
34. J. L. Paz, H. J. Franco, I. Reif, A. Marcano and F.García-Golding, *Phys. Rev. A* **37**, 3381 (1988).
35. J. L. Paz, A. E. Cárdenas and A. J. Hernández, *Quant. Opt.* **5**, 355 (1993).
36. J. L. Paz, M. García–Sucre, E. Squitieri and V. Mujica, *Chem. Phys. Lett.* **217**, 333 (1994).
37. M. Thoss, W. Miller and G. Stock, *J. Chem. Phys.* **112**, 10282 (2000).
38. T. Cusati, J. L. Paz, M. C. Salazar and A. J. Hernández, *Phys. Lett. A* **267**, 18 (2000).
39. J. L. Paz, T. Cusati, M. C. Salazar and A. J. Hernández, *J. Mol. Spect.* **211**, 198 (2002).
40. B. Di Bartolo, *Radiationless Processes* (Plenum Press, New York, 1980).

## Chapter 6

# Control of Molecular Chirality by Lasers

Kunihito Hoki and Yuichi Fujimura

*Department of Chemistry, Graduate School of Science,*
*Tohoku University, Sendai 980-8578, Japan*
*fujimurayuichi@mail.tains.tohoku.ac.jp*

Theoretical studies on laser control of molecular chirality, which have been carried out in our laboratory, are presented. Two fundamental issues about molecular chirality transformation in a racemic mixture are described. The first issue is that enantiomers should be pre-oriented to control chiral transformation by applying laser fields within the dipole approximation, and the second issue is that pure enantiomers can be selectively prepared from a pre-oriented racemic mixture by taking into account the photon polarization direction of the laser field. Based on these two issues, new control schemes for molecular chirality in a racemic mixture are presented. These control schemes include analytical treatment of preparation of enantiomers in the ground electronic state in an oriented racemic mixture, pump–dump control via an electronic excited state, control of molecular chirality in a randomly oriented racemic mixture using three polarization components of electric fields, stimulated Raman adiabatic passage method, and sequential pump–dump control of chirality transformation competing with photodissociation in an electronic excited state. For the three polarization components of laser fields, two polarization components are used for orientation of randomly oriented chiral molecules and the other component is used for chirality transformation of the oriented chiral molecules.

## 1. Introduction

Molecular chirality, or molecular handedness, is the characteristic of molecules having no inversion center or plane of symmetry. Synthesis of enantiomers, molecules with chirality, has been the main subject of interest in stereochemistry and biochemistry.[1]

For practical purposes, enantiomers are synthesized by use of catalysis to their chiral precursors.[2] In recent years, attention has been paid to selective preparation of enantiomers from their racemic mixture (equal mixture of two enantiomers).

With the development of laser science and technology together with the development of theoretical treatments, much interest has recently been shown in laser control of molecular reaction dynamics.[3–17] This is called coherent control or quantum control, in which the coherent nature of lasers is directly applied. Manipulation of molecular chirality by lasers is one of the fascinating targets in coherent control. Quantum control of molecular chirality is therefore expected to provide a new means for selective preparation of pure enantiomers from a racemic mixture.

Several types of quantum control of molecular chirality have been proposed. Shapiro and Brumer proposed a coherent control scenario for chiral molecular products from achiral precursors using achiral light, i.e., linearly polarized light.[18] Cina and Harris developed a wave packet theory and described the possibility of preparation and phase control of superposition states of $L$- and $R$-enantiomers in a symmetric, double-well potential via an electronically excited state, starting from a pure $L$- or $R$-enantiomer.[19] Duarte-Zamorano and Romero-Rochín analyzed Cina and Harris's theory by numerically solving the time-dependent Schrödinger equation.[20] Salam and Meath demonstrated the possibility of the control of excited state populations of enantiomers using circularly polarized pulses of various durations.[21] Shao and Hänggi presented a theory of absolute asymmetric synthesis using a circularly polarized field.[22] Shapiro *et al.* presented a theoretical method for coherently controlled asymmetric synthesis in a racemic mixture with achiral light.[23–26] Our group presented a quantum control theory for the selective preparation of enantiomers in pure state and mixed state cases, where the optimal control pulses were designed to prepare the pure enantiomers through a vibrationally excited level in the ground electronic state, and through an electronic excited state as well, in the ps time domain.[27–34] In this chapter, scenarios for laser control of molecular chirality, which have been developed in our laboratory, are described.

## 2. Fundamental Issues in Laser Control of Molecular Chirality

In this section, we consider two fundamental issues about molecular chiraity transformation in a racemic mixture by using electric field components of lasers.

### 2.1. *Laser control of an ensemble of racemic mixtures*

The first consideration is associated with the ensemble of racemic mixtures. Enantiomers in an isotropic ensemble, such as in gasses or in homogeneous solvents, cannot be controlled by applying electric fields of lasers within the dipole approximation. One possible way to control molecular chirality by applying laser fields within the dipole approximation is to fix the molecule in space. Although there is no difference between the energy structures of $R$- and $L$-forms except negligibly small parity-violating effects,[35–37] directions of vectors such as dipole moment vectors and transition moment vectors are different between them. Consider an achiral and isotropic ensemble whose density operator at the initial time$t = t_0$ is given as $\hat{\rho}(t_0)$. The symmetric initial density $\hat{\rho}(t_0)$ is invariant with respect to operation of the space inversion operator $\hat{P}$, i.e.,

$$\hat{P}\hat{\rho}(t_0)\hat{P} = \hat{\rho}(t_0). \tag{1}$$

Let $\hat{\rho}_{\mathrm{R}}$ be a density operator that produces the $R$-form from an ensemble by dipole moment interaction at a final time $t = t_{\mathrm{f}}$. The density operator is expressed in terms of the time evolution operator $\hat{U}(t_f, t_0; \boldsymbol{E})$ as

$$\hat{\rho}_R = \hat{U}(t_f, t_0; \boldsymbol{E})\hat{\rho}(t_0)\hat{U}^{\dagger}(t_f, t_0; \boldsymbol{E}). \tag{2}$$

Here $\boldsymbol{E}$, the electric field function of time, is explicitly denoted in the expression to emphasize that the time propagation of the density depends on the electric field through the dipole interaction. After operating $\hat{P}$ to $\hat{\rho}_{\mathbf{R}}$, we obtain[31]

$$\hat{\rho}_L = \hat{P}\hat{\rho}_R\hat{P} = \hat{U}(t_f, t_0; -\boldsymbol{E})\hat{\rho}(t_0)\hat{U}^{\dagger}(t_f, t_0; -\boldsymbol{E}). \tag{3}$$

It can be seen from Eqs. (2) and (3) that molecular chirality in a racemic mixture under an isotropic condition can be controlled if the initial phase of the laser field is properly taken into account. In this chapter, we omit discussion on any initial phase control of laser fields because there have been only a few exceptional case studies on an initial phase dependence of femtosecond laser fields to control molecular chirality.[38] This means that it is extremely difficult to create a chiral density from an achiral density only by taking into account the dipole interaction. Therefore, we focus on a pre-oriented racemic mixture in this review article. We also describe how to orient a chiral system by using lasers.

### 2.2. *Photon polarizations of lasers*

It is well recognized that circularly polarized electric fields of UV or visible lasers can predominantly decompose one enantiomeric constituent of chiral molecules in a racemic mixture in solution. The origin of the asymmetric destruction is the difference in molecular absorption coefficients between the two enantiomers. This difference is due to a simultaneous contribution of the electronic and magnetic dipole moments. Existence of circularly polarized electric fields is one of the possible mechanisms of homochirality in life on the earth: absolute asymmetric synthesis of the same kind of enantiomers of amino acids in a racemic mixture was performed under the condition of circularly polarized radiation created by neutron stars in the universe. However, it is insufficient to produce an appreciable enantiomeric excess using lasers in ordinary experimental conditions. This is because the magnetic field interaction is very weak in comparison to the electric dipole interaction. Therefore, we consider molecular chirality control by using only the electric field component of lasers.

Let us now consider an ensemble of pre-oriented enantiomers in a racemic mixture, i.e., an equal mixture of right (R)-handed and left (L)-handed chiral molecules. We will selectively prepare L-handed enantiomers from the racemic mixture by using lasers.[33] For simplicity, we assume that the chiral molecule is characterized by a symmetric, one-dimensional double-well potential in the electronic ground state as shown in Fig. 1. The stable configuration of the R-handed enantiomer and that of the L-handed

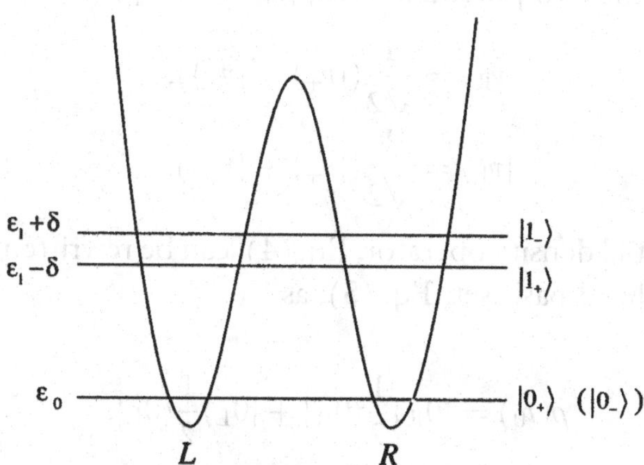


**Fig. 1.** One-dimensional double-well potential in the electronic ground state of a chiral molecule. L and R denote the positions of two enantiomers, L-form and R-form, respectively. $\varepsilon_0$ is an approximate eigenvalue of the doubly degenerate states, $|0_+\rangle(|0_-\rangle)$ because of a high potential energy barrier. $\varepsilon_1+\delta$ and $\varepsilon_1-\delta$ are the eigenvalues of the first and second excited states, respectively. The energy separation between these two excited states is given as $2\delta$.[33] Reproduced with permission from American Chemical Society.

one are denoted by R and L, respectively. The lowest vibrational state, denoted by $|0_+\rangle$, and the first excited state, $|0_-\rangle$, are degenerated with energy $\varepsilon_0$ because of its high potential energy barrier. Suffix plus (minus) in the vibrational eigenstates represents symmetric (anti-symmetric) with respect to the chiral plane that divides enantiomers between L- and R-forms. Let the other two excited states be the second excited state denoted by $|1_+\rangle$ with energy $\varepsilon_1 - \delta$ and the third excited state $|1_-\rangle$ with energy $\varepsilon_1 + \delta$. Here 2 $\delta$ denotes the energy separation between them.

Assuming that the energy difference between $|0_\pm\rangle$ and $|1_\pm\rangle$ is much larger than $kT$, where $k$ is the Boltzmann constant and $T$ is the temperature, the initial density operator $\hat{\rho}(t_0)$ is expressed in the low temperature limit as

$$\hat{\rho}\,(t_0) = |0_+\rangle\,\frac{1}{2}\,\langle 0_+| + |0_-\rangle\,\frac{1}{2}\,\langle 0_-| \tag{4}$$

Alternatively, the system can be expressed in terms of the localized states into the two potential wells, $|v_R\rangle$ and $|v_L\rangle$ ($v = 0$ or $1$) as

$$|v_R\rangle = \frac{1}{\sqrt{2}}(|v_+\rangle - |v_-\rangle), \tag{5a}$$

$$|v_L\rangle = \frac{1}{\sqrt{2}}(|v_+\rangle + |v_-\rangle). \tag{5b}$$

The initial density operator, Eq. (4), can be rewritten in terms of the localized basis set, Eq. (5), as

$$\hat{\rho}(t_0) = |0_R\rangle\frac{1}{2}\langle 0_R| + |0_L\rangle\frac{1}{2}\langle 0_L|. \tag{6}$$

We now specify the target operator for controlling pure enantiomers from a racemic mixture. As mentioned in the previous section, the eigenvalues of $\hat{\rho}(t)$, which are statistical weight, are invariant when the time-propagation of $\hat{\rho}(t)$ is a unitary process. Therefore, the final state is also expressed in terms of a one-to-one statistical mixture.

Let the target population be localized in one of the wells, e.g., to produce the R-form of an enantiomer. In this case, the target operator is given as

$$\hat{W} = |0_R\rangle \frac{1}{2} \langle 0_R| + |1_R\rangle \frac{1}{2} \langle 1_R|. \tag{7}$$

The Hamiltonian of the total system, $\hat{H}(t)$, is expressed within the semiclassical treatment of the radiation interaction with matter as

$$\hat{H}(t) = \hat{H}_0 - \boldsymbol{\mu} \cdot \boldsymbol{E}(t). \tag{8}$$

Here, $\hat{H}_0$ is the molecular Hamiltonian, $\boldsymbol{\mu}$ is the dipole moment vector, and $\boldsymbol{E}(t)$ is the electric field of the laser used:

$$\boldsymbol{E}(t) = 2\boldsymbol{A}(t) \cos \omega t, \tag{9}$$

where $\boldsymbol{A}(t)$ is the pulse envelope with a photon-polarization vector, and $\omega$ is the carrier frequency.

The Hamiltonian matrix, $\boldsymbol{H}(t)$, is given in the eigenstate representation as

$$\boldsymbol{H}(t) = \begin{pmatrix} \varepsilon_0 & 0 & -\langle 0_+|\boldsymbol{\mu}|1_+\rangle \cdot \boldsymbol{E}(t) & -\langle 0_+|\boldsymbol{\mu}|1_-\rangle \cdot \boldsymbol{E}(t) \\ 0 & \varepsilon_0 & -\langle 0_-|\boldsymbol{\mu}|1_+\rangle \cdot \boldsymbol{E}(t) & -\langle 0_-|\boldsymbol{\mu}|1_-\rangle \cdot \boldsymbol{E}(t) \\ -\langle 1_+|\boldsymbol{\mu}|0_+\rangle \cdot \boldsymbol{E}(t) & -\langle 1_+|\boldsymbol{\mu}|0_-\rangle \cdot \boldsymbol{E}(t) & \varepsilon_1 - \delta & 0 \\ -\langle 1_-|\boldsymbol{\mu}|0_+\rangle \cdot \boldsymbol{E}(t) & -\langle 1_-|\boldsymbol{\mu}|0_-\rangle \cdot \boldsymbol{E}(t) & 0 & \varepsilon_1 + \delta \end{pmatrix}. \tag{10}$$

Here, terms such as $\langle 1_+|\boldsymbol{\mu}|0_+\rangle$ denote the matrix elements of dipole moment operator $\boldsymbol{\mu}$.

The following relation between the dipole matrix elements

$$\langle 1_+|\boldsymbol{\mu}|0_+\rangle = \langle 1_-|\boldsymbol{\mu}|0_-\rangle \quad \text{and} \quad \langle 1_+|\boldsymbol{\mu}|0_-\rangle = \langle 1_-|\boldsymbol{\mu}|0_+\rangle, \tag{11}$$

can easily be seen since the matrix elements in the localized basis set are expressed in a good approximation as

$$\langle 1_L|\boldsymbol{\mu}|0_R\rangle = 0 \quad \text{and} \quad \langle 1_R|\boldsymbol{\mu}|0_L\rangle = 0. \tag{12}$$

Ability of selective preparation of enantiomers from a pre-oriented racemic mixture depends on the photon polarization direction of the laser field that brings about asymmetry in the racemic mixture. This can easily be seen from Eq. (11). L-enantiomers are optically active between the ground state $|0_L\rangle$ and excited states $|1_\pm\rangle$, while R-enantiomers in the ground state are inactive if the polarization of the laser is set as

$$-\langle 1_+|\boldsymbol{\mu}|0_+\rangle \cdot \boldsymbol{A}(t) = -\langle 1_+|\boldsymbol{\mu}|0_-\rangle \cdot \boldsymbol{A}(t) \equiv \hbar\Omega(t). \tag{13}$$

The above expression is equivalent to

$$-\langle 1_+|\boldsymbol{\mu}|0_R\rangle\cdot \boldsymbol{A}(t) = 0 \quad \text{and} \quad -\langle 1_+|\boldsymbol{\mu}|0_L\rangle\cdot \boldsymbol{A}(t) = -\sqrt{2}\hbar\Omega(t), \tag{14a}$$

$$-\langle 1_-|\boldsymbol{\mu}|0_R\rangle\cdot \boldsymbol{A}(t) = 0 \quad \text{and} \quad -\langle 1_-|\boldsymbol{\mu}|0_L\rangle\cdot \boldsymbol{A}(t) = -\sqrt{2}\hbar\Omega(t), \tag{14b}$$

or

$$-\langle 1_R|\boldsymbol{\mu}|0_R\rangle\cdot \boldsymbol{A}(t) = 0 \quad \text{and} \quad -\langle 1_L|\boldsymbol{\mu}|0_L\rangle\cdot \boldsymbol{A}(t) = -2\hbar\Omega(t). \tag{14c}$$

If we use an electric field of a linearly polarized laser whose polarization direction is determined by Eq. (13), we can see that the population of R-enantiomers, $P_R(t)$, is expressed as

$$P_R(t) = \langle 0_R|\hat{\rho}(t)|0_R\rangle + \langle 1_R|\hat{\rho}(t)|1_R\rangle \geq \frac{1}{2}. \tag{15}$$

The population of R-enantiomers in the excited state $|1_R\rangle$, $\langle 1_R|\hat{\rho}(t)|1_R\rangle$, is transferred from L-enantiomers in the ground state $\langle 0_L|\hat{\rho}(t_0)|0_L\rangle = 1/2$ via a tunneling process after optical excitation.

Similarly, if we set the laser field polarization as

$$\langle 1_+|\boldsymbol{\mu}|0_+\rangle \cdot \boldsymbol{A}'(t) = -\langle 1_+|\boldsymbol{\mu}|0_-\rangle \cdot \boldsymbol{A}'(t) \equiv \hbar\Omega'(t), \tag{16}$$

then we obtain

$$\langle 1_{\rm L}|\boldsymbol{\mu}|0_{\rm L}\rangle\cdot \boldsymbol{A}'(t) = 0 \quad \text{and} \quad \langle 1_{\rm R}|\boldsymbol{\mu}|0_{\rm R}\rangle\cdot \boldsymbol{A}'(t) = 2\hbar\Omega'(t). \tag{17}$$

Thus, the total population of L-enantiomers, $P_{\rm L}(t)$, is increased compared with that in the initial racemic mixture. This is the qualitative explanation for the laser control of enantiomers from its pre-oriented racemic mixture based on the basic principle of the control scenario described above.

### 2.3. *Density matrix treatment of a racemic mixture*

Based on the two fundamental issues described above, we present laser control of molecular chirality in a racemic mixture in a quantitative way by using the simple model system shown in the previous sub-section. For this purpose, we derive an analytical expression of the time development of populations of enantiomers under the laser field condition, Eq. (13) in the dressed state representation. The time evolution of the density operator $\hat{\rho}(t)$ obeys the Liouville equation

$$i\hbar\frac{d}{dt}\hat{\rho}(t) = [\hat{H}(t), \hat{\rho}(t)], \tag{18}$$

where [ , ] is a commutator.

For simplicity, we restrict ourselves to enantiomer control by using a stationary laser field. The equation of motion of the density matrix $\rho(t)$ represented by field-free eigenstates is

$$i\hbar\frac{d}{dt}\rho(t) = \left[\begin{pmatrix} \varepsilon_0 & 0 & 2\hbar\Omega\cos(\omega t) & 2\hbar\Omega\cos(\omega t) \\ 0 & \varepsilon_0 & 2\hbar\Omega\cos(\omega t) & 2\hbar\Omega\cos(\omega t) \\ 2\hbar\Omega\cos(\omega t) & 2\hbar\Omega\cos(\omega t) & \varepsilon_1 - \delta & 0 \\ 2\hbar\Omega\cos(\omega t) & 2\hbar\Omega\cos(\omega t) & 0 & \varepsilon_1 + \delta \end{pmatrix}, \rho(t)\right] \tag{19}$$

with the initial condition

$$\rho(t_0) = \begin{pmatrix} 1/2 & 0 & 0 & 0 \\ 0 & 1/2 & 0 & 0 \\ 0 & 0 & 0 & 0 \\ 0 & 0 & 0 & 0 \end{pmatrix}. \tag{20}$$

In Eq. (19), Rabi frequency $\Omega$ is defined as $\hbar\Omega = -\langle 1_+|\boldsymbol{\mu}|0_+\rangle \cdot \boldsymbol{A}$, where $\boldsymbol{A}$ is the amplitude of the laser field with a polarization.

Within the rotating wave approximation, Eq. (19) is rewritten as

$$i\hbar\frac{d}{dt}\tilde{\rho}(t) = \left[ \begin{pmatrix} 0 & 0 & 0 & 0 \\ 0 & 0 & 0 & 0 \\ 0 & 0 & -\hbar\tilde{\Omega} & 0 \\ 0 & 0 & 0 & \hbar\tilde{\Omega} \end{pmatrix}, \tilde{\rho}(t) \right], \tag{21}$$

where

$$\tilde{\rho}(t) = \boldsymbol{\Lambda}(t)\boldsymbol{R}(t)\rho(t)\boldsymbol{R}^{-1}(t)\boldsymbol{\Lambda}^{-1}(t) \tag{22}$$

with

$$\boldsymbol{R}(t) = \begin{pmatrix} \exp(i\varepsilon_0 t/\hbar) & 0 & 0 & 0 \\ 0 & \exp(i\varepsilon_0 t/\hbar) & 0 & 0 \\ 0 & 0 & \exp(i\varepsilon_1 t/\hbar) & 0 \\ 0 & 0 & 0 & \exp(i\varepsilon_1 t/\hbar) \end{pmatrix} \tag{23}$$

and

$$\boldsymbol{\Lambda}(t)^{-1} = \begin{pmatrix} \frac{1}{\sqrt{2}} & \frac{\delta}{\sqrt{2}\hbar\tilde{\Omega}} & \frac{\Omega}{\tilde{\Omega}} & \frac{\Omega}{\tilde{\Omega}} \\ -\frac{1}{\sqrt{2}} & \frac{\delta}{\sqrt{2}\hbar\tilde{\Omega}} & \frac{\Omega}{\tilde{\Omega}} & \frac{\Omega}{\tilde{\Omega}} \\ 0 & \frac{\sqrt{2}\Omega}{\tilde{\Omega}} & \frac{-\delta-\hbar\tilde{\Omega}}{2\hbar\tilde{\Omega}} & \frac{-\delta+\hbar\tilde{\Omega}}{2\hbar\tilde{\Omega}} \\ 0 & \frac{\sqrt{2}\Omega}{\tilde{\Omega}} & \frac{\delta-\hbar\tilde{\Omega}}{2\hbar\tilde{\Omega}} & \frac{\delta+\hbar\tilde{\Omega}}{2\hbar\tilde{\Omega}} \end{pmatrix}. \tag{24}$$

Here,

$$\tilde{\Omega} = \sqrt{\delta^2/\hbar^2 + 4\Omega^2}. \tag{25}$$

The solution of the equation of motion, Eq. (21), is obtained as

$$\tilde{\rho}(t) = \begin{pmatrix} 1 & 0 & 0 & 0 \\ 0 & 1 & 0 & 0 \\ 0 & 0 & \exp[i\tilde{\Omega}(t-t_0)] & 0 \\ 0 & 0 & 0 & \exp[-i\tilde{\Omega}(t-t_0)] \end{pmatrix} \times \tilde{\rho}(t_0) \begin{pmatrix} 1 & 0 & 0 & 0 \\ 0 & 1 & 0 & 0 \\ 0 & 0 & \exp[-i\tilde{\Omega}(t-t_0)] & 0 \\ 0 & 0 & 0 & \exp[i\tilde{\Omega}(t-t_0)] \end{pmatrix} \tag{26}$$

with $\tilde{\rho}(t_0) = \rho(t_0)$.

Therefore, the time development of the population in each localized state can be expressed in an analytical form as

$$P_{0\mathrm{R}} = \langle 0_\mathrm{R}|\hat{\rho}(t)|0_\mathrm{R}\rangle = \frac{1}{2}, \tag{27a}$$

$$P_{0\mathrm{L}} = \langle 0_\mathrm{L}|\hat{\rho}(t)|0_\mathrm{L}\rangle = \frac{1}{2}\left[\frac{\delta^2 + 4\hbar^2\Omega^2\cos(\tilde{\Omega}t)}{\hbar^2\tilde{\Omega}^2}\right]^2, \tag{27b}$$

$$P_{1\mathrm{R}} = \langle 1_\mathrm{R}|\hat{\rho}(t)|1_\mathrm{R}\rangle = 2\left[\frac{\delta\Omega[1-\cos(\tilde{\Omega}t)]}{\hbar\tilde{\Omega}^2}\right]^2, \tag{27c}$$

and

$$P_{1\mathrm{L}} = \langle 1_\mathrm{L}|\hat{\rho}(t)|1_\mathrm{L}\rangle = 2\left[\frac{\Omega\sin(\tilde{\Omega}t)}{\tilde{\Omega}}\right]^2. \tag{27d}$$

Equation (27d) indicates that in a stationary laser field, the population in $|1_L\rangle$ is maximized at $\tilde{\Omega}t = \pi/2$ and the population of $|1_R\rangle$ transferred by tunneling from $|1_L\rangle$ is maximized at $\tilde{\Omega}t = \pi$. The maximum population in the localized state, $|1_R\rangle$, is obtained if Rabi frequency satisfies $|\Omega| = \delta/(2\hbar)$. The total population in the right-hand well is expressed as

$$P_R = \frac{1}{2} + 2\left(\frac{\Omega \sin\left(\tilde{\Omega}t\right)^2}{\tilde{\Omega}}\right). \tag{28}$$

We now consider $H_2POSH$ with axial chirality as a chiral molecule shown in Fig. 2 to demonstrate how enantiomers in a pre-oriented racemic mixture are selectively prepared by applying a linearly polarized laser whose polarization direction is properly determined. The reaction path is along the torsional coordinate of the SH bond around the PS bond of $H_2POSH$. Let $H_2POSH$ be pre-oriented in such a way that the PS bond is fixed in the $X$–$Z$ plane. The torsional potential function is characterized by a double well potential similar to that in Fig. 1. An *ab initio*

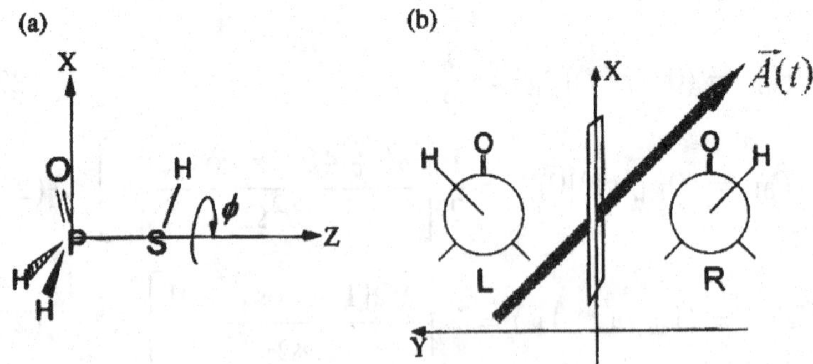


**Fig. 2.** (a) Reaction coordinate $\phi$ of axial chirality control of $H_2POSH$ in space-fixed Cartesian coordinates $(X, Y, Z)$. $\phi$ is the torsional coordinate of the SH bond around the PS bond. (b) Two enantiomers, L-form and R-form, are located at $\phi = -60$ degrees and 60 degrees, respectively. $A(t)$ denotes the polarization vector of the electric field of laser pulse to control L- to R-enantiomers. The angle of $A$ with respect to the $X$-axis is 58 degrees, which is approximately parallel to the S–H bond of the R-enantiomer.[33] Reproduced with permission from American Chemical Society.

molecular orbital calculation shows that the L-form enantiomer is located at the torsional angle of −60 degrees and that the R-form is located at 60 degrees. Its barrier height is 500 $cm^{-1}$. We adopt the four-state model described above to control chirality of $H_2POSH$: this model consists of the torsional ground, first, second and third excited states. The condition of quasi-degeneration between the ground and first excited states is valid within a time scale of a few hundred ps since the level splitting between them is only 0.053 $cm^{-1}$. This corresponds to tunneling time of 630 ps. The level splitting between the second and third excited states, $2\delta$, is 1.6 $cm^{-1}$, which corresponds to tunneling time of 21 ps. The dipole components $\mu_X$ and $\mu_Y$, which are, respectively, symmetric and anti-symmetric with respect to the $ZX$-plane involving the O, P, and S atoms, were taken into account in determining the direction of the photon polarization from Eq. (13). From $|\langle 1_+|\boldsymbol{\mu}|0_L\rangle| = 0.19$ D, which was estimated by using an *ab initio* MO method, the direction of photon polarization with respect to the $X$-axis was estimated to be 58 degrees. This direction is almost parallel to the SH-bond of $H_2POSH$ in the R-form as shown in Fig. 2.

Figure 3 shows the time evolution of the chirality transformation in a racemic mixture under an optimal condition in which Rabi frequency satisfies the condition $|\Omega| = \delta/(2\hbar)$, i.e., 100% of R-enantiomer are produced from the racemic mixture. In this ideal case, both the Rabi oscillation and tunneling rate are synchronously matched, and all of the population of $|1_L\rangle$ created from the ground state is transferred to the target state $|1_R\rangle$ by a tunneling process. The tunneling rate depends on the laser intensity applied.

## 3. Control Scenarios

In this section, several scenarios for control of molecular chirality are presented.

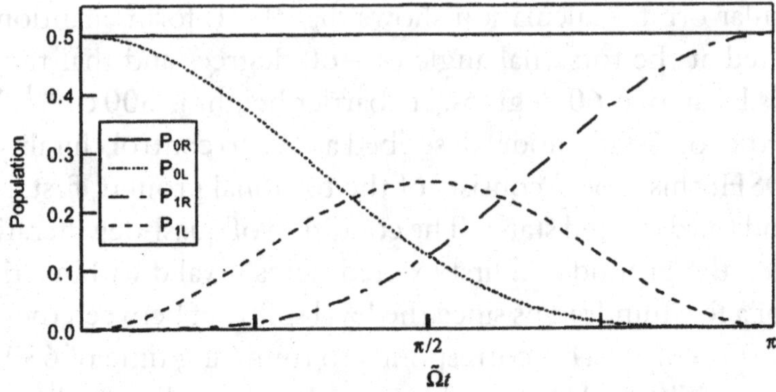


**Fig. 3.** Time evolution of the chirality transformation of L- to R-enantiomers in a racemic mixture in an optimal control condition. $P_{0\mathrm{R}}(P_{0\mathrm{L}})$ denotes the population of the R- (L-) enantiomer in the lowest ground state $|0_\mathrm{R}\rangle(|0_\mathrm{L}\rangle)$. $P_{1\mathrm{R}}(P_{1\mathrm{L}})$ denotes the population of the R- (L-) enantiomer in the lowest ground state $|1_\mathrm{R}\rangle(|1_\mathrm{L}\rangle)$. $P_{0\mathrm{R}}$ is independent of time. The chiral transformation takes place via $|1_\mathrm{L}\rangle$ from $|0_\mathrm{L}\rangle$ by an optimal pulse and then from $|1_\mathrm{L}\rangle$to the final state, $|1_\mathrm{R}\rangle$, by tunneling.[33] Reproduced with permission from American Chemical Society.

### 3.1. *Pump–dump control via an electronic excited state*

The pump–dump method is the simplest method for laser control of chemical reactions. In this subsection, an application of pump–dump method to chirality control of pre-oriented chiral molecules in a racemic mixture is presented.[34] A femosecond pump–dump control via an electronic excited state is considered, as shown in Fig. 4. In this figure, the potential energy function in the ground state is characterized by a double well potential with its chiral nature. $|g0_\mathrm{R}\rangle(|g0_\mathrm{L}\rangle)$ and $|gn_\mathrm{R}\rangle(|gn_\mathrm{L}\rangle)$ denote the lowest ground state with energy $\varepsilon_{g0}$ and $n$th vibrational state with energy $\varepsilon_{gn}$ in the ground electronic state of the R- (L-) enantiomer, respectively. The potential energy function in the electronic excited state, on the other hand, is characterized by a bound potential with a single minimum well, which means that the stable nuclear configuration in the excited electronic state is achiral, not chiral. $|em\rangle$ denotes the vibronic state with energy $\varepsilon_{em}$ in the excited electronic state. We

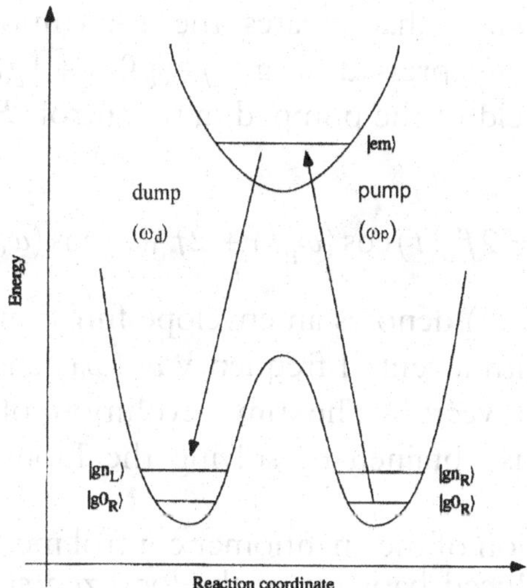


**Fig. 4.** A scheme of a femtosecond pump–dump pulse control for molecular chirality of R- to L-enantiomers via an electronic excited state with a vibronic state $|em\rangle$. The potential energy in the ground state is characterized by a double-well potential whose minimum positions corresponding to the L- and R-enantiomers. The potential in the electronic excited state is a bound potential with a single minimum well. $\omega_p$ ($\omega_d$) denotes the central frequency of the pump (dump) pulse.

notice that there is a difference in the direction of the transition moments between two enantiomers, L-form and R-form, although the magnitudes are equal. For example, for an optical transition from the ground state $|g0\rangle$ to the vibronic state $|em\rangle$, $\langle em|\mu|g0_L\rangle \neq \langle em|\mu|g0_R\rangle$ and $|\langle em|\mu|g0_L\rangle| = |\langle em|\mu|g0_R\rangle|$. Here $\langle em|\mu|g0_L\rangle$ and $\langle em|\mu|g0_R\rangle$ are matrix elements of the dipole transition moment operator between $|em\rangle$ and the lowest ground states of L- and R-enantiomers, respectively.

Consider a chirality control from R-enantiomers to L-enantiomers in which the energy difference between $|g0_\pm\rangle$ and $|g1_\pm\rangle$ is much larger than $kT$. For a low temperature limit, the initial density operator $\hat{\rho}(t_0)$ is expressed as

$$\hat{\rho}(t_0) = |\,g0_R\rangle\frac{1}{2}\langle\,g0_R| + |\,g0_L\rangle\frac{1}{2}\langle\,g0_L|. \qquad (29)$$

The target operator that creates the L-enantiomers from a racemic mixture is expressed as $|g0_L\rangle\frac{1}{2}\langle g0_L| + |gn_L\rangle\frac{1}{2}\langle gn_L|$.

An electric field of the pump–dump control, $\boldsymbol{E}(t)$, is simply given as

$$\boldsymbol{E}(t) = 2f_p(t)\cos(\omega_p t) + 2f_d(t)\cos(\omega_d t), \tag{30}$$

where $f_p(t)$ ($f_d(t)$) denotes an envelope function of the pump (dump) pulse with a central frequency $\omega_p(\omega_d)$ and with a photon polarization vector. The time evolution of the racemic mixture, $\hat{\rho}(t_0)$, is obtained by solving the Liouville equation, Eq. (18).

For explanation of the enantiomer control mechanism, let us consider the reduced basis set of the localized states shown in Fig. 4. The Hamiltonian matrix, $\boldsymbol{H}(t)$, is given in the localized basis set representation as

$$\boldsymbol{H}(t) = \begin{pmatrix} \varepsilon_{g0} & 0 & 0 & 0 & -\langle g0_L|\boldsymbol{\mu}|em\rangle\cdot\boldsymbol{E}(t) \\ 0 & \varepsilon_{g0} & 0 & 0 & -\langle g0_R|\boldsymbol{\mu}|em\rangle\cdot\boldsymbol{E}(t) \\ 0 & 0 & \varepsilon_{gn} & 0 & -\langle gn_L|\boldsymbol{\mu}|em\rangle\cdot\boldsymbol{E}(t) \\ 0 & 0 & 0 & \varepsilon_{gn} & -\langle gn_R|\boldsymbol{\mu}|em\rangle\cdot\boldsymbol{E}(t) \\ -\langle em|\boldsymbol{\mu}|g0_L\rangle\cdot\boldsymbol{E}(t) & -\langle em|\boldsymbol{\mu}|g0_R\rangle\cdot\boldsymbol{E}(t) & -\langle em|\boldsymbol{\mu}|gn_L\rangle\cdot\boldsymbol{E}(t) & -\langle em|\boldsymbol{\mu}|gn_R\rangle\cdot\boldsymbol{E}(t) & \varepsilon_{em} \end{pmatrix}. \tag{31}$$

For simplicity, consider a pump–dump control scheme in which the two pulses are separated and do not temporarily overlap with each other.[24] In this case, an analytical expression for the time evolution of the system can be derived. The time evolution of the molecular system can be divided into two sequential processes: one is the process in which the population is transfered in the excited vibronic state, i.e., achiral state, by applying the pump pulse, and the other is the dump process in which the population is transferred from the excited state to the final target state by applying the dump pulse.

First consider the pump process through which the population of the R-handed enantiomers is transferred from the initial ground state to the excited vibronic state, while the population of the L-handed molecules in the lowest vibrational state remains unchanged. For this process, the pump pulse has to satisfy the condition described in Sec. 2.2 as given below:

$$\langle em|\boldsymbol{\mu}|\mathrm{g}0_{\mathrm{L}}\rangle \cdot \boldsymbol{f}_{\mathrm{p}}(t) = 0, \tag{32a}$$

and

$$\langle em|\boldsymbol{\mu}|\mathrm{g}0_{\mathrm{R}}\rangle \cdot \boldsymbol{f}_{\mathrm{p}}(t) = -\hbar\Omega_{\mathrm{p}}(t). \tag{32b}$$

Here, $|\Omega_{\mathrm{p}}(t)|$ is the so-called Rabi frequency of the pump pulse. Equation (32a) indicates that L-handed enantiomers in the vibronic state can be prepared by using the linearly polarized pump pulse in such a way that the polarization is orthogonal with respect to the direction of the transition moment of the enantiomers.

The time evolution of the racemic mixture in the pump process is obtained by solving the equation of motion of the density matrix $\rho(t)$ in terms of the reduced field-free set of the dominant states, $|\mathrm{g}0_{\mathrm{R}}\rangle$, $|\mathrm{g}0_{\mathrm{L}}\rangle$, and $|em\rangle$, as

$$i\hbar\frac{d}{dt}\rho(t) = \left[\begin{pmatrix} \varepsilon_{\mathrm{g}0} & 0 & 0 \\ 0 & \varepsilon_{\mathrm{g}0} & 2\hbar\Omega_{\mathrm{p}}(t)\cos(\omega_{\mathrm{p}}t) \\ 0 & 2\hbar\Omega_{\mathrm{p}}(t)\cos(\omega_{\mathrm{p}}t) & \varepsilon_{em} \end{pmatrix}, \rho(t)\right] \tag{33}$$

with the initial condition

$$\rho(t_0) = \begin{pmatrix} 1/2 & 0 & 0 \\ 0 & 1/2 & 0 \\ 0 & 0 & 0 \end{pmatrix}. \tag{34}$$

The central frequency of the pump laser is given as $\hbar\omega_{\mathrm{p}} = \varepsilon_{em} - \varepsilon_{g0}$ under the resonant condition. The equation of motion, Eq. (33), is transformed within the rotating wave approximation as

$$i\hbar\frac{d}{dt}\tilde{\rho}(t) = \left[\begin{pmatrix} 0 & 0 & 0 \\ 0 & 0 & \hbar\Omega_{\mathrm{p}}(t) \\ 0 & \hbar\Omega_{\mathrm{p}}(t) & 0 \end{pmatrix}, \tilde{\rho}(t)\right], \tag{35}$$

where

$$\tilde{\rho}(t) = \boldsymbol{R}(t)\boldsymbol{\rho}(t)\boldsymbol{R}^{-1}(t). \tag{36}$$

Matrix $\boldsymbol{R}(t)$ in Eq. (36) is given as

$$\boldsymbol{R}(t) = \begin{pmatrix} \exp\left\{\dfrac{i\varepsilon_{g0}(t-t_0)}{\hbar}\right\} & 0 & 0 \\ 0 & \exp\left\{\dfrac{i\varepsilon_{g0}(t-t_0)}{\hbar}\right\} & 0 \\ 0 & 0 & \exp\left\{\dfrac{i\varepsilon_{em}(t-t_0)}{\hbar}\right\} \end{pmatrix}. \tag{37}$$

The solution of the equation of motion, Eq. (35), is obtained as

$$\boldsymbol{\rho}(t) = \boldsymbol{R}^{-1}(t)\boldsymbol{U}(t,t_0)\boldsymbol{\rho}(t_0)\boldsymbol{U}(t_0,t)\boldsymbol{R}(t) \tag{38}$$

with $\tilde{\boldsymbol{\rho}}(t_0) = \boldsymbol{\rho}(t_0)$. Here,

$$U(t,t_0) = \begin{pmatrix} 1 & 0 & 0 \\ 0 & \cos\dfrac{\phi_{\mathrm{p}}(t,t_0)}{2} & -i\sin\dfrac{\phi_{\mathrm{p}}(t,t_0)}{2} \\ 0 & -i\sin\dfrac{\phi_{\mathrm{p}}(t,t_0)}{2} & \cos\dfrac{\phi_{\mathrm{p}}(t,t_0)}{2} \end{pmatrix}, \tag{39}$$

where $\phi_{\rm p}(t, t_0)$, the pulse area of the pump laser, is defined as

$$\phi_{\rm p}(t, t_0) = 2\int_{t_0}^{t} \Omega_{\rm p}(t')dt'. \tag{40}$$

The diagonal density matrix elements are expressed in an analytical form as

$$\rho_{\rm g0_L,g0_L}(t) = \frac{1}{2}, \tag{41a}$$

$$\rho_{\rm g0_R,g0_R}(t) = \frac{1}{2}\cos^2\frac{\phi_{\rm p}(t, t_0)}{2}, \tag{41b}$$

and

$$\rho_{em,em}(t) = \frac{1}{2}\sin^2\frac{\phi_{\rm p}(t, t_0)}{2}. \tag{41c}$$

We can see from Eqs. (41) that the population is completely transferred to the vibronic state $|em\rangle$ from the lowest vibrational ground state $|{\rm g0_R}\rangle$ for $\phi_{\rm p}(t, t_0) = \pi$, i.e., by using a $\pi$ pump pulse.

Let us consider the dump process to create the L-handed enantiomers in the vibrationally excited ground state $|{\rm g}n_{\rm L}\rangle$ from $|em\rangle$. The initial population in the dump process is given by Eq. (41c). The polarization vector of the dump pulse to control L-handed enantiomers is determined by the conditions

$$\langle {\rm g}n_{\rm R}|\boldsymbol{\mu}|em\rangle \cdot f_{\rm d}(t) = 0 \tag{42a}$$

and

$$\langle {\rm g}n_{\rm L}|\boldsymbol{\mu}|em\rangle \cdot f_{\rm d}(t) = -\hbar\Omega_{\rm d}(t). \tag{42b}$$

Here, $|\Omega_{\rm d}(t)|$ is the Rabi frequency in the dump process.

By using the same procedure as that described for the pump process, we can derive an analytical expression for the population in the target state $|gn_L\rangle$ and that in the vibronic state at $t_f$ as

$$\rho_{gn_L,gn_L}(t_f) = \frac{1}{2}\sin^2\frac{\phi_d(t_f,t)}{2}\sin^2\frac{\phi_p(t,t_0)}{2} \tag{43a}$$

and

$$\rho_{em,em}(t_f) = \frac{1}{2}\cos^2\frac{\phi_d(t_f,t)}{2}\sin^2\frac{\phi_p(t,t_0)}{2}. \tag{43b}$$

Here, $\phi_d(t_f,t) = 2\int_t^{t_f}\Omega_d(\tau)d\tau$. The resonant condition $\hbar\omega_d = \varepsilon_{em} - \varepsilon_{gn}$ and the rotation wave approximation were used in deriving Eq. (43). The yield of the L-handed enantiomers $Y_L(t_f)$ can then be expressed as

$$\begin{aligned} Y_L(t) &= \langle g0_L|\rho(t_f)|g0_L\rangle + \langle gn_L|\rho(t_f)|gn_L\rangle \\ &= \frac{1}{2}\left\{1+\sin^2\frac{\phi_d(t_f,t)}{2}\sin^2\frac{\phi_p(t,t_0)}{2}\right\}. \end{aligned} \tag{44}$$

The maximum population can be obtained by applying both $\pi$ pump and dump pulses to the racemic mixture.

So far, a pump–dump laser control scheme to create L-handed enantiomers from a racemic mixture has been described. In a similar way, R-handed enantiomers can also be created from a racemic mixture.

We now apply the pump–dump control method to $H_2POSH$, one of the simplest real systems possessing axial chirality. The torsional motion of the SH bond around the PS bond is the reaction coordinate defining the chirality. Both the electronic ground and singlet excited state potential energy curves of the same types as those sketched in Fig. 1 and transition dipole moments have been calculated using time-dependent density functional theory (B3LYP/6-311+G(2p,d) level of theory).[41] We prepare L-handed enantiomers from a racemic mixture in the low temperature limit. We set $|g0_L\rangle\frac{1}{2}\langle g0_L| + |g1_L\rangle\frac{1}{2}\langle g1_L|$ as the target operator. The tunneling time defined as the whole oscillation period R→L→R in the first vibrationally excited doublet is estimated to be 34 ps.[41] In other words, control of the molecular chirality must be completed within this time. The vibronic state of fifth torsional quantum number $m = 5$, $|e5\rangle$ of the bound vibronic states in the first singlet excited state was adopted as the intermediate excited state because of the strongest transition moment from the lowest ground state.[41] The energy differences between $|e5\rangle$ and the adjacent vibronic states, $|e4\rangle$ and $|e6\rangle$, are 339 cm$^{-1}$ and 217 cm$^{-1}$, respectively. The transition energy from the lowest ground state to $|e5\rangle$ is 44,252 cm$^{-1}$. The transition energy between $|e5\rangle$ and the target state, $|g1_L\rangle$, is 44,067 cm$^{-1}$. The energy difference between $|g0_L\rangle(|g0_R\rangle)$ and $|g1_L\rangle(|g1_R\rangle)$ is 185 cm$^{-1}$.

The direction of the polarization vector that is determined by Eqs. (31) and (41) to create a population in the excited state from $|g0_R\rangle$ is 45 degrees with respect to the P–S bond of $H_2POSH$ in the case in which the laser propagates along the $X$-axis. Then, the polarization vector with the envelope function $f_p(t)$ of the pump pulse was set to be of the following form:

$$f_p(t) = \begin{cases} (\boldsymbol{e}_Y + \boldsymbol{e}_Z) E_p \sin^2(\alpha_p t) & (\text{for} \quad 0 \leq t \leq 300 \text{ fs}), \\ 0 & (\text{for } 300 \text{ fs} < t). \end{cases} \tag{45a}$$

Here, $\boldsymbol{e}_Y$ and $\boldsymbol{e}_Z$ are the unit vectors of $Y$ and $Z$ directions in the space-fixed coordinates, respectively, $7.4 \times 10^9$ V/m was used as the amplitude of the pulse $E_p$, and $\pi/300$ fs$^{-1}$ was used as the width parameter of the envelope function, $\alpha_p$.

In a similar way, the direction of the polarization vector of the dump pulse is −45 degrees to the P–S bond of $H_2POSH$ in the same laser setup as that used for the pump laser. The polarization vector with the envelope function $f_d(t)$ of the dump pulse was set to be of the following form:

$$f_p(t) = \begin{cases} (\boldsymbol{e}_Y - \boldsymbol{e}_Z) E_d \sin^2\{\alpha_d(t - 300)\} & \\ \qquad (\text{for } 300 \leq t \leq 900 \text{ fs}), & \\ 0 \qquad (\text{for } t < 300 \text{ fs}, 900 \text{ fs} < t). & \end{cases} \tag{45b}$$

Here, $2.2 \times 10^9\,\text{V}/m$ was used as the amplitude of the dump pulse $E_{\text{d}}$ and $\alpha_{\text{d}} = \pi/600\,\text{fs}^{-1}$ was used as the width parameter of the dump pulse.

The bandwidth of the pump pulse and that of the dump pulse were estimated to be 50 and $25\,\text{cm}^{-1}$, respectively, by using Fourier transformation of Eqs. (45a) and (45b). These bandwidths are narrow compared with the level spacings, $\sim 200\,\text{cm}^{-1}$ in the ground state and those, $\sim 300\,\text{cm}^{-1}$ in the resonant intermediate electronic state. Therefore, the pump and dump control method can be applied almost perfectly within the approximate five-state model.

Figure 5(a) and 5(b) show the time evolution of the wave packets in the lowest torsional state and that in the first excited quantum state in the electronic ground state in the presence of pump and dump pulses. Here, complementary to the five states described above, a convergent set of other eigenstates was taken into account in order to quantitatively evaluate the dynamics. We can see from Fig. 5 how effectively the molecular chirality is controlled by applying the pump and dump pulses.

Figure 6 shows the population changes of the localized states as a function of time in a pump–dump laser control scheme. We can see a high yield of L-handed enantiomers, $Y_L(t_f) = 0.96$ at the final time $t_f = 800\,\text{fs}$.

To demonstrate the effectiveness of the new method, we have exemplarily considered the control of dynamic chirality of $H_2POSH$. The present results will serve as a reference for extended studies including additional degrees of freedom and competing processes as shown in Sec. 3.4.

### 3.2. *Control of molecular chirality in a randomly oriented racemic mixture using three polarization components of electric fields*

Chiral molecules have so far been assumed to be oriented or attached to a surface because the control of molecular chirality

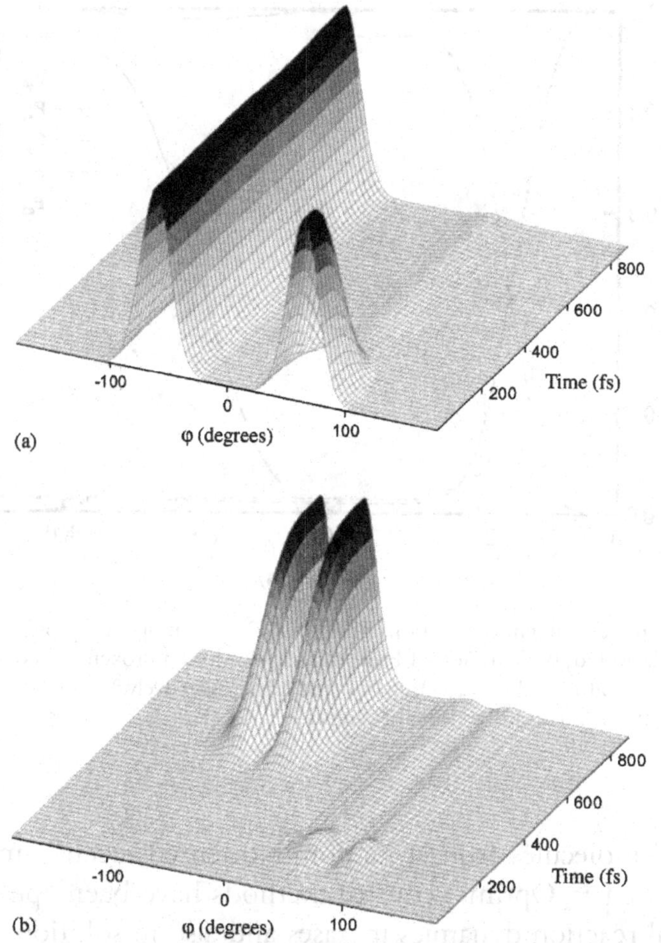


**Fig. 5.** Time evolution of wavepackets in chirality control from R-handed to L-handed enantiomers of $H_2POSH$ in a racemic mixture by a pump–dump pulse method. (a) Wavepackets of $|g0_R\rangle$ and $|g0_L\rangle$. (b) Wavepackets of $|g1_R\rangle$ and $|g1_L\rangle$.[34] Reproduced with permission from American Institute of Physics.

was performed by using the dipole interaction term of laser fields. The electric field components of lasers propagating toward a spatially fixed direction alone do not make any contribution to the production of enantiomeric excesses from a randomly oriented racemic mixture. In this subsection, an optimal control scenario

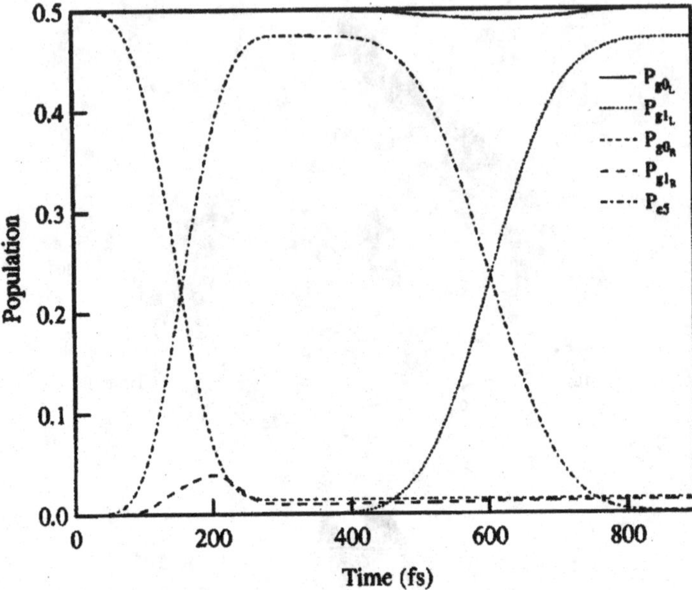


**Fig. 6.** Time-dependent populations of $H_2POSH$ in a pump–dump pulse control. Here, solid, dotted, broken, dotted-broken, and two-dotted broken lines denote the populations of $|g0_L\rangle$, $|g1_L\rangle$, $|g0_R\rangle$, $|g1_R\rangle$, and $|e5\rangle$, respectively.[34] Reproduced with permission from American Institute of Physics.

of chiral molecules from a randomly oriented racemic mixture is presented.[42] Optimal control methods have been applied to chemical reaction dynamics in gases and also in solution. Here, this is based on utilization of the electric field components of laser pulses, not the magnetic field component. The interaction between enantiomers and electric fields is taken into account within the electric dipole and the long wave approximation. The main control idea is to use two kinds of lasers: i.e., UV pulse laser and IR pulse laser. Firstly, an IR laser pulse is prepared as a useful tool for orienting molecules in a homogeneously distributed system and, secondly, electric field components of UV pulses produce a sufficient enantiomeric excess in an ultra-short time scale when an oriented racemic mixture is prepared.

For this purpose, three pulsed lasers whose electric fields are $E_X(t)$, $E_Y(t)$, and $E_Z(t)$ are used. Here, suffies $X$, $Y$, and $Z$ of $E(t)$ denote the polarization directions of the electric fields of lasers in the space-fixed coordinates. One laser, $E_Z(t)$, is used for orienting enantiomers, and the other two, $E_X(t)$ and $E_Y(t)$, are used for preparing pure enantiomers from the pre-oriented ones.

To demonstrate this sequential control scenario, let us adopt phosphinotioic acid, $H_2POSH$, as a realistic model again. The torsion of the SH bond around the PS bond is the reaction coordinate of the preparation of pure enantiomers. We employ a simplified treatment of a three-dimensional model in which the system consists of a rotational degree of freedom $\theta$, the P–S stretching vibrational mode $R$, and reaction coordinate $\phi$ as shown in Fig. 7. The rotation was assumed to be quasi-diatomic. The stretching mode was assumed to be a harmonic one. The constants associated with the rotation and the vibration are estimated using an *ab initio* MO method.

The total Hamiltonian $H(t)$ is given within the semi-classical treatment of radiation–matter interactions as

$$H(t) = H_0 + V_Z(t) + V_X(t) + V_Y(t). \tag{46}$$

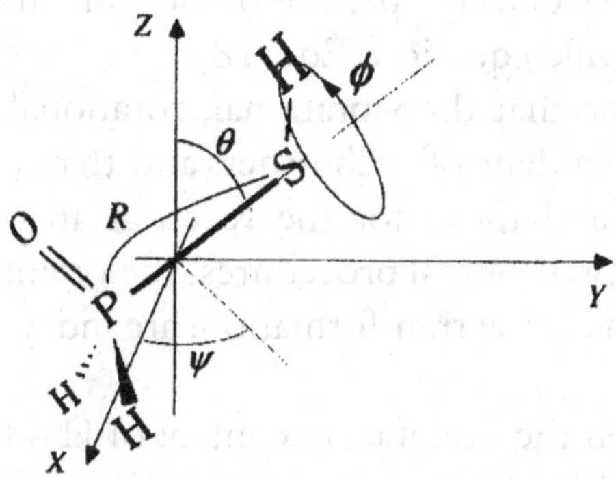


**Fig. 7.** A configuration of $H_2POSH$ in space-fixed Cartesian coordinates $(X, Y, Z)$. Angle $\theta$ denotes the orientation angle, $R$ is the vibrational mode of the PS bond, and $\phi$ is the reaction coordinate of molecular chirality of a randomly oriented $H_2POSH$.[42] Reproduced with permission from American Institute of Physics.

The zero-order Hamiltonian $H_0$ is expressed as a sum of rotational Hamiltonian $H_0(\theta, \psi)$, vibrational Hamiltionian $H_0(R)$, and chiral Hamiltonian, $H_0(\phi)$, as

$$H_0 = H_0(\theta, \psi) + H_0(R) + H_0(\phi), \tag{47}$$

where each Hamiltonian is assumed to be independent of each other.

In Eq. (46), $V_Z(t)$ is the interaction Hamiltonian between $H_2POSH$ and the IR laser to induce the orientation, and $V_X(t)$ and $V_Y(t)$ are the interaction Hamiltonians, preparing pure enantiomers from the pre-oriented racemic mixture by UV laser pulses. These are expressed as

$$\begin{aligned} V_Z(t) &= -\mu_Z(R)E_Z(t)\cos\theta \\ &\quad -\frac{1}{2}[\alpha_{\parallel}(R)\cos^2\theta + \alpha_{\perp}(R)\sin^2\theta]E_Z^2(t), \quad (48a) \\ V_X(t) &= -\mu_X(\phi)E_X(t), \quad (48b) \\ V_Y(t) &= -\mu_Y(\phi)E_Y(t). \quad (48c) \end{aligned}$$

Here, $\mu_Z(R)$, $\mu_X(R)$ and $\mu_Y(R)$ are $Z$, $X$, and $Y$ components of the dipole moment vector, and $\alpha_{\parallel}(R)$ and $\alpha_{\perp}(R)$ are parallel and perpendicular components of the polarizability, respectively.

The rovibronic density operator of the chiral molecules, $\rho(t)$, satisfies the Louville equation, Eq. (18).

Let us assume that the vibrational, rotational, and reaction modes are independent of each other, and that laser modes for the orientation and those for the reaction are different from each other, i.e., two control procedures, the orientation of chiral molecules and the chiral transformation, are independent of each other.

We first go on the orientation control of $H_2POSH$.[43] Here, we consider a value of cos$\theta$ in the range $0 \leq \theta \leq \pi$ as an extent of orientation (see Fig. 7). Thus, the control is to maximize $\mathrm{Tr}\{\rho(t_f)\cos\theta\}$ under the condition of the minimum input of the applied laser. That is, the optimal laser pulse for orientation can theoretically be designed by maximizing the objective functional $O(E_Z)$ as

$$O(E_Z) = \mathrm{Tr}\{\rho(t_f)\cos\theta\} - \frac{1}{\hbar A}\int_{t_0}^{t_f} dt\, E_Z(t)^4, \qquad (49)$$

where $\rho(t_f)$ denotes the density operator at final time $t_f$, which depends on the electric field of the laser applied $E_Z(t)$, $A$ is a weight-factor, and the fourth power of $E_Z(t)$ is adopted for fast convergence of the optimal procedure.

We are interested in nonadiabatic orientation control in ps time regimes, i.e., control paths involving pure rotational transitions are omitted in the orientation control. In the actual control procedure, the electric field of the pulse was assumed to be given as

$$E_Z(t) = \sum_{n=n_{\min}}^{\infty} a_n \cos\frac{n\pi t}{t_f} + b_n \sin\frac{n\pi t}{t_f}, \qquad (50)$$

where $n_{\min}$ is set to take a minimum allowed frequency. The objective functional $O(E_Z)$ was optimized with respect to coefficients $a_n$ and $b_n$ in Eq. (50).

Figure 8 shows the results of optimal control of orientation of $H_2POSH$ starting from a randomly oriented racemic mixture in the low temperature limit. In Fig. 8(a), the temporal behavior of $\langle\cos\theta\rangle$ is shown as a measure of the magnitude of the orientation. The orientation was set to be carried out within 1 ps. From Fig. 8(a), we can see that the orientation is completed at the final time.

Figure 8(b) shows the amplitudes of the optimal laser pulse for the orientation. Figure 8(c) shows the frequency-resolved spectrum of the optimal pulse in which the strongest intensity band around 1400 cm$^{-1}$ corresponds to the fundamental frequency of the stretching vibration of the PS bond of $H_2POSH$.

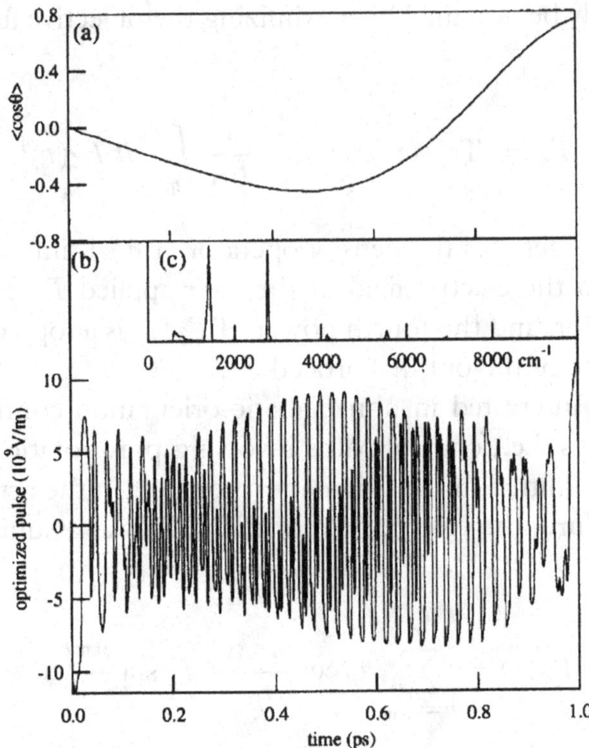


**Fig. 8.** (a) Temporal behaviors of the orientation of $H_2POSH$ in a randomly oriented racemic mixture under the optimal laser field condition. (b) Optimal laser pulse shape for the orientation control. (c) Frequency-resolved spectrum of the electric field of the optimal pulse. Two main bands around 1400 and 280 $cm^{-1}$ are the fundamental and overtone of the stretching vibration of the PS bond, respectively.[42] Reproduced with permission from American Institute of Physics.

The coherent excitation of the fundamental and overtone vibrational modes of the stretching vibration is the main origin for creation of the orientation. Two excitations are, respectively, induced by dipole and polarizability interactions between the molecule and radiation field of the perturbation in Eq. (48a). The former transition is of g–u and the latter is of g–g from the symmetry argument. The interaction between these two transitions creates asymmetry of the system, i.e., the orientation shown in Fig. 8(a).

The other oriented configuration that is opposite to that derived above can be prepared by applying the laser pulse whose electric field is given as $-E_Z(t)$.

The electric field of the pulses can generally be expressed as that of a simple, multicolor laser pulse as

$$E_Z(t) = \sum_n c_n \sin[n(\omega t + \delta_n)] \tag{51a}$$

and

$$-E_Z(t) = \sum_n c_n \sin[n(\omega t + \delta_n^+ \pi/n)], \tag{51b}$$

where $\omega$ is the fundamental frequency, $c_n$ is the coefficient of each component, and $\delta_n$ is the phase of each frequency component. Equation (51) shows that the phase difference of each frequency component is different between $E_Z(t)$ and $-E_Z(t)$. This indicates that the orientation direction is determined by the phase difference, i.e., $\delta_n - \delta_{n'}$ for one orientation, and $(\delta_n+\pi/n) - (\delta_{n'}+\pi/n')$ for the opposite orientation. The orientation control described above is based on breaking of the symmetry by the relative phase of the different pulse components.

We now consider creation of $L$-enantiomers from the preoriented $H_2POSH$ racemic mixture by using linearly polarized UV lasers. For simplicity, the chiral Hamiltonian, $H_0(\phi)$, was assumed to be expressed within a five-state model as

$$\begin{aligned} H_0(\phi) = {} & |0_R\rangle\varepsilon_0\langle 0_R| + |0_L\rangle\varepsilon_0\langle 0_L| + |1_R\rangle \\ & \times \varepsilon_1\langle 1_R| + |1_L\rangle\varepsilon_1\langle 1_L| + |e\rangle\varepsilon_e\langle e|, \end{aligned} \tag{52}$$

where $|n_R\rangle$ and $|n_L\rangle$ denote R- and L-enantiomer states with torsional vibrational level $n$, and $|e\rangle$ denotes a vibronic eigenstate of an electronically excited state. $\varepsilon_0$, $\varepsilon_1$, and $\varepsilon_e$ are the energies. The initial density operator $\rho(t_0)$ for the racemic mixture is written in the low temperature limit as

$$\rho(t_0) = |0_R\rangle\frac{1}{2}\langle 0_R| + |0_L\rangle\frac{1}{2}\langle 0_L|. \tag{53}$$

We note that the oriented $H_2POSH$ prepared by laser $E_Z(t)$ still has freedom of rotation around the $Z$-axis. We select two, rotation-fixed configurations: one with $X$, $Y$, and $Z$ and the other with $-X$, $-Y$, and $Z$, i.e., the $\pi$-rotated one. Chirality control of L-enantiomers from the racemic mixture of $H_2POSH$ with the two configurations can be carried out by applying the pump–dump method described in the previous section: first a laser polarization direction of the pump laser pulse is chosen as

$$\langle 0_L|(\pm\mu_X, \pm\mu_Y)\cdot\begin{pmatrix} E_X(t) \\ E_Y(t) \end{pmatrix}|e\rangle \neq 0, \tag{54a}$$

and

$$\langle 0_R|(\pm\mu_X, \pm\mu_Y)\cdot\begin{pmatrix} E_X(t) \\ E_Y(t) \end{pmatrix}|e\rangle = 0. \tag{54b}$$

Here, signs $\pm$ correspond to the two configurations.

When a $\pi$ pulse whose polarization direction is determined by Eqs. (54) is used under the resonant excitation condition, density operator at time $t_1\,\rho(t_1)$ just after the excitation is expressed as

$$\rho(t_1) = |0_R\rangle\frac{1}{2}\langle 0_R| + |e\rangle\frac{1}{2}\langle e|. \tag{55}$$

Figure 9 shows temporal behaviors of the populations of the zeroth order states in a pump and dump control scheme. Intensities of the pump and dump pulses are of $1.10 \times 10^{10}$ and $4.0\times10^9$ V/m, respectively. The duration of each pulse was taken to be 2 ps. The directions of the polarization were at angles of 6 and $-6$ degree with respect to the $X$-axis for the pump and dump processes, respectively. The control results show a high yield of $P_R(t_f) = 0.77$ at $t_f = 2$ ps. The time duration of 2 ps is sufficiently long compared with a rotational period of $\sim$ 16 ps.

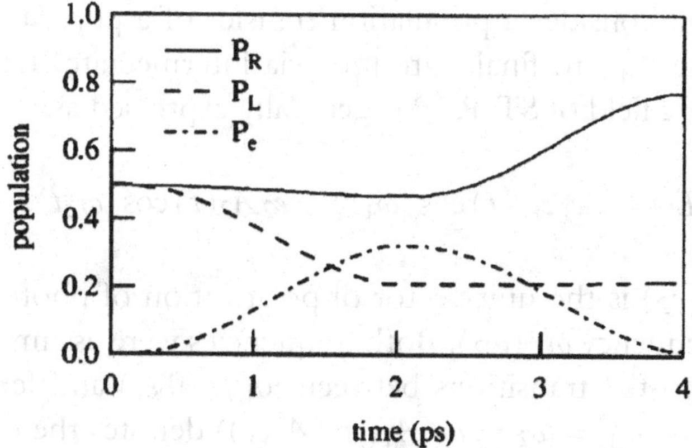


**Fig. 9.** Temporal behaviors of the populations of a randomly oriented racemic mixture of $H_2POSH$ in a pump–dump pulse control scheme. A line denotes the population of R-handed enantiomers, $P_R$, a broken line denotes that of L-enantiomers, $P_L$, and a dotted broken line denotes that of the electronic excited state, $P_e$.[42] Reproduced with permission from American Institute of Physics.

Therefore, the enantiomer control can be performed under the pre-oriented condition.

### 3.3. *Stimulated Raman adiabatic passage method*

Stimulated Raman adiabatic passage (STIRAP) is one of the methods for complete population transfer.[44–46] This method is carried out by applying the Stokes pulse before the pump laser counterintuitively. The feature of STIRAP is that it is robust and no sophisticated experimental setup is needed compared to other laser control methods such as pulse shaping techniques. STIRAP has been applied mostly to non-degenerated systems in which the pump and Stokes processes do not overlap each other in the frequency domain, because the pump and Stokes pulses cannot be distinguished. Therefore, we cannot directly apply STIRAP to control of molecular chirality that belongs to a degenerate system. In this subsection, we present a new version of STIRAP that is applicable to a degenerate system.[47]

We first consider a population transfer of a pure state from initial state $|0_L\rangle$ to final state $|0_R\rangle$ via intermediate state $|em\rangle$. The electric field of STIRAP is generally expressed as

$$E(t) = e_1 A_1(t)\cos(\omega_1 t) + e_2 A_2(t)\cos(\omega_2 t), \qquad (56)$$

where $e_1(e_2)$ is the unit vector of polarization of photons with central frequency $\omega_1$ ($\omega_2$). Both frequencies were assumed to be resonant to the transitions between $|0_L\rangle$ ($|0_R\rangle$) and $|em\rangle$, and simplified as $\omega_1 = \omega_2 = \omega$. $A_1(t)$ ($A_2(t)$) denotes the envelope of the laser pulse 1(2).

Equation (56) can be rewritten as

$$\boldsymbol{E}(t) = \frac{\boldsymbol{A}(t)}{|\boldsymbol{A}(t)|}|\boldsymbol{A}(t)|\cos(\omega t) = \boldsymbol{\eta}(t)A(t)\cos(\omega t), \qquad (57)$$

where $\boldsymbol{A}(t) = e_1 A_1(t) + e_2 A_2(t)$, $A(t) \equiv |\boldsymbol{A}(t)|$, and

$$\boldsymbol{\eta}(t) \equiv \boldsymbol{A}(t)/|\boldsymbol{A}(t)|. \qquad (58)$$

Equation (58), the time-dependent unit vector of the polarization direction, indicates that two lasers with the same frequency and two different linearly polarizations are equivalent to a single laser with a time-dependent polarization unit vector.

The chiral molecular Hamiltonian in the interaction picture can be expressed within the rotating wave approximation (RWA) under a resonance condition as

$$H_I(t) = -\frac{\boldsymbol{\eta}(t)A(t)}{2}\cdot(|0_L\rangle\boldsymbol{\mu}_{0L,em}\langle em| + |0_R\rangle\boldsymbol{\mu}_{0R,em}\langle em|) + \text{h.c.}, \qquad (59)$$

where $\mu_{0L,em}$ ($\mu_{0R,em}$) denotes the transition dipole moment in the **L**(**R**) enantiomer. If the relative angle between $\boldsymbol{\eta}(t)$ and $\boldsymbol{\mu}_{0L,em}$ ($\mu_{0L,em}$) is defined as $\eta_L(t)$ ($\eta_R(t)$), Eq. (59) can be rewritten as

$$H_I(t) = -\frac{A(t)}{2}\mu_{0,em}(\cos\eta_{\rm L}(t)|0_{\rm L}\rangle\langle em| + \cos\eta_{\rm R}(t)|0_{\rm L}\rangle\langle em|) + \text{h.c.} \tag{60}$$

Here, $\mu_{0,em}$ is the absolute value of $\mu_{0{\rm L},em}(\mu_{0{\rm R},em})$.

The eigenvalues and eigenfunctions of the Hamiltonian can be obtained as

$$E_- = \frac{\hbar\sqrt{\Omega_1^2(t)+\Omega_2^2(t)}}{2}, \tag{61a}$$

$$|u_-\rangle = \frac{1}{\sqrt{2}}(\sin\Theta|0_{\rm L}\rangle - |em\rangle + \cos\Theta|0_{\rm R}\rangle), \tag{61b}$$

$$E_0 = 0, \tag{61c}$$

$$|u_0\rangle = \cos\Theta|0_{\rm L}\rangle - \sin\Theta|0_{\rm R}\rangle, \tag{61d}$$

$$E_+ = -\frac{\hbar\sqrt{\Omega_1^2(t)+\Omega_2^2(t)}}{2}, \tag{61e}$$

and

$$|u_+\rangle = \frac{1}{\sqrt{2}}(\sin\Theta|0_{\rm L}\rangle + |em\rangle + \cos\Theta|0_{\rm R}\rangle), \tag{61f}$$

with

$$\Theta(t) = \tan^{-1}\frac{\Omega_1(t)}{\Omega_2(t)}. \tag{62}$$

Here, $\Omega_1(t)$ and $\Omega_2(t)$, time-dependent Rabi frequencies, are defined as

$$\Omega_1(t) = \frac{\mu_{0,em}\cos\eta_{\rm L}(t)A(t)}{\hbar} \tag{63a}$$

and

$$\Omega_2(t) = \frac{\mu_{0,em}\cos\eta_{\rm R}(t)A(t)}{\hbar}, \tag{63b}$$

respectively. These dressed states have the same structures as those of a conventional STIRAP. It is understood from Eqs. (63) that complete population transfer in a degenerate system can be achieved by taking into account only the adiabatic change in the polarization direction of the linearly polarized electric field. The population $P_{0R}(t)$ of $|0_R\rangle$ due to transfer from $|0_L\rangle$ is expressed as

$$P_{0R}(t) = \sin^2 \Theta(t), \tag{64}$$

where

$$\Theta(t) = \tan^{-1} \frac{\cos \eta_L(t)}{\cos \eta_R(t)}. \tag{65}$$

Complete population transfer can be performed by setting the adiabatic changes both in $\eta_L(t)$ and $\eta_R(t)$ from the initial time $t = 0$ to the final time $t = t_f$ as

$$\begin{aligned} \eta_L(0) = \frac{\pi}{2} \quad &\text{and} \quad \eta_R(0) \neq \frac{\pi}{2}; \\ \eta_L(t_f) \neq \frac{\pi}{2} \quad &\text{and} \quad \eta_R(t_f) = \frac{\pi}{2}. \end{aligned} \tag{66}$$

It is convenient to define the relative angle between two transition moment vectors $\boldsymbol{\mu}_{0L,em}$ and $\boldsymbol{\mu}_{0R,em}$ as $\alpha(0 \leq \alpha \leq \pi)$. Assuming that the unit vector $\boldsymbol{\eta}(t)$ evolves in the plane formed by the two transition moment vectors, we obtain

$$\alpha = \eta_L(t) - \eta_R(t). \tag{67}$$

Figure 10 shows a two-dimensional geometrical structure of the time-dependent $\boldsymbol{\eta}(t)$. A complete population transfer from $|0L\rangle$ to $|0R\rangle$ is performed by letting $\eta_R(t)$ be swept as

$$\eta_R(0) = \frac{\pi}{2} - \alpha \rightarrow \eta_R(t_f) = \frac{\pi}{2} \quad (\text{for } 0 \leq \alpha \leq \pi/2) \tag{68a}$$

and

$$\eta_R(0) = \frac{3\pi}{2} - \alpha \rightarrow \eta_R(t_f) = \frac{\pi}{2} \quad (\text{for } \pi/2 \leq \alpha \leq \pi) \tag{68b}$$

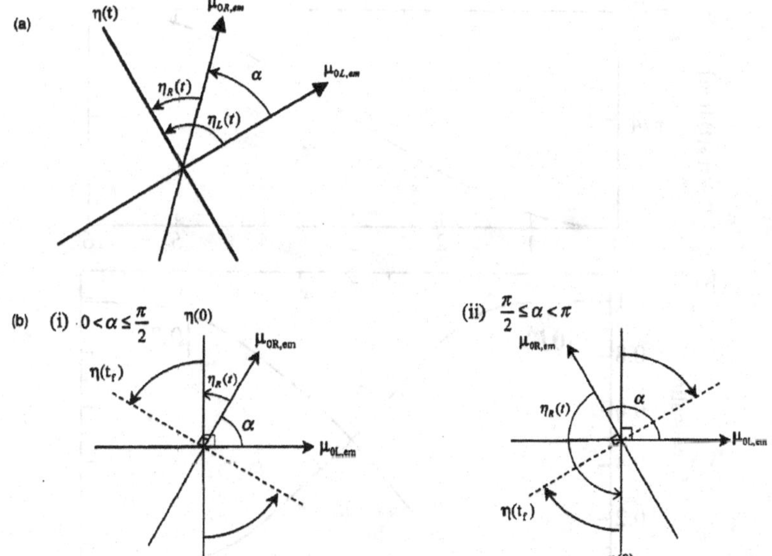

**Fig. 10.** (a) Two-dimesional geometrical structure of time-dependent polarization vector $\eta(t)$, where $\alpha$ is the angle between two transition moments, $\mu_{0L,em}$ and $\mu_{0R,em}(0 \leq \alpha \leq \pi)$. $\eta_L(t)$ $(\eta_R(t))$ is the relative angle between $\eta(t)$ and $\mu_{0L,em}$ $(\mu_{0R,em})$. (b) The possible direction of the photon polarization vector $\eta(t)$ for $|0L\rangle \rightarrow |0R\rangle$ transfer in the pure state case: (i) $0 \leq \alpha \leq \pi/2$ and (ii) $\pi/2 \leq \pi$.[47] Reproduced with permission from American Institute of Physics.

Figure 11 shows the results of application of the new STIRAP method to control of molecular chirality of $H_2POSH$ in a pure state case. A three-state model with the initial state $|0_L\rangle$, the final state $|0_R\rangle$, and the fourth vibrational excited state in the first electronic excited state, $|e5\rangle$, was employed. Here, the electric field of the laser pulse applied has the form

$$\boldsymbol{E}(t) = \boldsymbol{\eta}(t) A_0 \cos \omega t, \tag{68}$$

where

$$A_0 = 2.2 \times 10^9 \, \text{V/m}, \quad \eta_L(t) = \frac{\pi}{2} + \frac{\pi}{2} \sin^2\left(\frac{\pi}{2}\frac{t}{t_f}\right), \quad \text{and}$$
$$\eta_R(t) = \eta_L(t) - \tfrac{\pi}{2} \quad \text{with } t_f = 6.0\,\text{ps}.$$

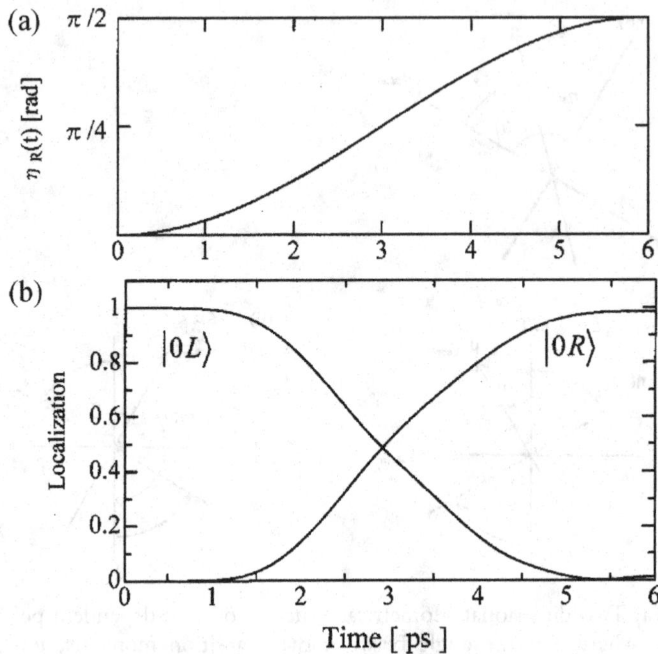


**Fig. 11.** Stimulated Raman adiabatic passage (STIRAP) control of molecular chirality from $|0_L\rangle$ to $|0_R\rangle$ of $H_2POSH$ in a pure state case within a three-state model. (a) Form of the STIRAP variable $\eta_R(t)$; (b) temporal behavior of populations of the two states, $|0_L\rangle$ and $|0_R\rangle$.[47] Reproduced with permission from American Institute of Physics.

It can be seen from Fig. 11 that molecular chirality is nearly completely transferred within the final time. The robustness of the solutions with respect to the form of $\eta_L(t)$ ($\eta_R(t)$) and parameters $A_0$ and $t_f$ was confirmed.

A new STIRAP method for control of molecular chirality in the pure state case has been presented. The STIRAP method is also applicable to control of molecular chirality in a mixed case.[47]

### 3.4. *Sequential pump–dump control of chirality transformation competing with photodissociation in an electronic excited state*

We have so far restricted our chirality control scenario to a one-dimensional system, omitting dynamic effects such as dissociations and dephasings. It was found from *ab ititio* MO calculations that the first excited state of the chiral molecule $H_2POSH$ or $H_2POSD$ is characterized by a repulsive potential energy surface along the P–S bond.[48] In this subsection, a new type of pump–dump control method,[48] consisting of two sequential pump–dump series of pulses, is proposed in order to control chiral transformation of $H_2POSD$ with a repulsive potential in its electronic excited state (see Fig. 12). This method has

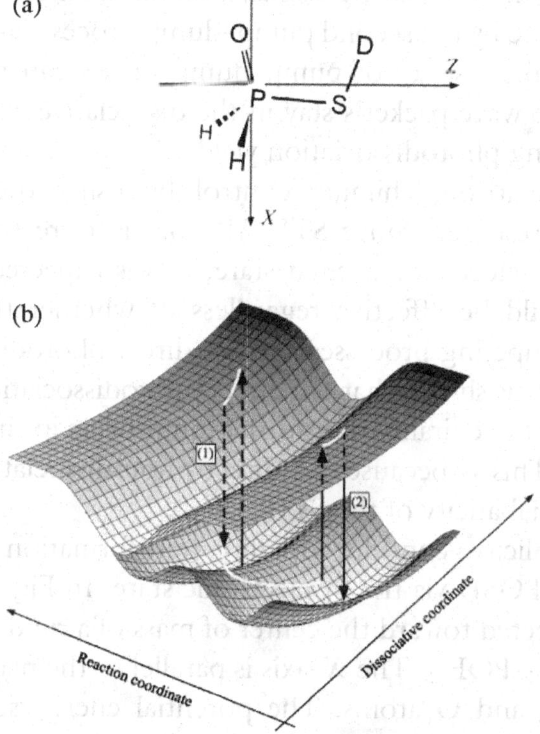


**Fig. 12.** (a) Configuration of an axial chiral molecule, $H_2POSD$, together with $X$ and $Z$ axes. The molecule has two stable configurations, in which the dihedral angle O–P–S–D is about −60 and 60 degrees. (b) Two sequential pump–dump pulse control scenario designed by the optimal control theory. The arrows denoted by (1) correspond to the first pump–dump process, and the arrows denoted by (2) correspond to the second pump–dump process.[48] Reproduced with permission from American Chemical Society.

been derived by applying an optimal control theory with an iteration algorithm. The designed scenario consists of application of sequential pump-dump pulses as indicated by the arrows in Fig. 12(b). A nuclear wave packet with sufficient kinetic energy to cross the potential barrier in the electronic ground state is prepared via the excited state by the first pump–dump process (denoted by arrows (1) in Fig. 12(b)). The wave packet runs from the potential well to the other one on the ground state. The wave packet having a large amount of kinetic energy is then transformed into one having a small amount of kinetic energy via an excited state by the second pump–dump process (arrows (2)).

As a result, a series of pump–dump pulses can reduce the period of the wave packet's stay in the dissociative excited state for minimizing photodissociation yields.

For comparison, chirality control by using the STIRAP method is presented. Since STIRAP does not create any population in an electronic excited state, it was expected that this method would be effective regardless of whether the excited state has competing processes such as direct photodissociation. However, it was shown that the direct photodissociation process prevails over the chirality transformation process in the STIRAP procedure. This is because the direct photodissociation breaks down the adiabaticity of the STIRAP.

For simplicity, consider chirality transformation of a pre-oriented $H_2POSD$ via the $S_1$ electronic state. In Fig. 12(a), the $Z$-axis is directed toward the center of mass of a rigid part, S–D, from that of $-POH_2$. The $X$-axis is parallel to the plane formed by the S, P, and O atoms. The potential energy surfaces are expressed in terms of two nuclear coordinates, $\phi$ and $r$. Here, $\phi$, the reaction coordinate of the chirality transformation, is the dihedral angle of O–P–S–D defining molecular chirality (R) or (L). The coordinate $r$ represents the distance between the center of mass of a rigid part, $-POH_2$, and that of –SD, and its potential in the $S_1$ state is repulsive. Other degrees of vibrational freedom were considered uncoupled and, therefore, frozen spectator modes.

By using these two variables $\phi$ and $r$, the system Hamiltonian is written as

$$H(\phi, r, t) = \begin{pmatrix} h_g(\phi, r) & -\mu_{ge}(\phi, r) \cdot E(t) \\ -\mu_{eg}(\phi, r) \cdot E(t) & h_e(\phi, r) \end{pmatrix}, \quad (69)$$

where the vibrational Hamiltonian of each electronic state is given by

$$h_s(\phi, r) = -\frac{\hbar^2}{2m_r}\frac{\partial^2}{\partial r^2} - \frac{\hbar^2}{2m_\phi}\frac{\partial^2}{\partial \phi^2} + V_s(\phi, r). \quad (70)$$

Here, the subscript $s$ stands for g or e. The potential energy surfaces $V_g(\phi, r)$ and $V_e(\phi, r)$ and transition dipole memont vector $\mu_{ge}(\phi, r)$ were computed using an *ab inito* MO method.[41]

Before showing the results obtained by optimal control theory, we present the results obtained by using the STIRAP method described in the previous subsection. Let us consider the chirality transformation from the L-form to the R-form of $H_2POSD$. Figure 13 shows the results obtained by using the STIRAP method. The chirality transformation yield is only 0.02%, and the photodissociation probability is 0.93 at $t = 20$ ps. These results mean that the chirality transformation and dissociation processes are competing processes. STIRAP does not create any population in the intermediate state if the adiabatic condition is satisfied. However, the dissociation process causes breakdown of the adiabatic condition, and most of the wave packet created in the $S_1$ state goes into the direct photodissociation channel. To determine the origin of the breakdown of adiabatic condition, we applied the STIRAP method to a model system of $H_2POSD$ in which the nuclear motion along $r$ is fixed at $r = 2.35$ Å to prevent a direct photodissociation process. The results are shown in Fig. 13(b). The same electric field as that used in Fig. 13(a) was

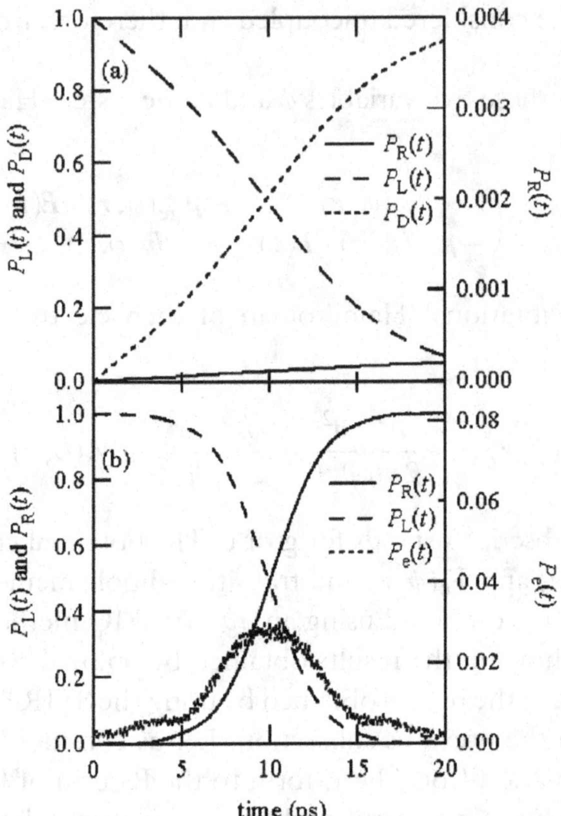


**Fig. 13.** Time-dependent populations of $H_2$POSD by STIRAP. (a) Results for the dissociation process included as a competing processs. $P_R(t)$ and $P_L(t)$ are populations localizing at right and left potential wells in the $S_0$ state, respectively. $P_D(t)$ is the sum of the population in the electric excited state and that absorbed by the boundary, which corresponds to the dissociation probability. (b) Results for chirality control omitting the dissociation process. $P_e(t)$ is a population in the $S_1$ state.[48] Reproduced with permission from American Chemical Society.

applied. The chirality transformation process from (L) to (R) is completed with about 100% yield.

It should be noted that an attempt to make the passage time shorter to suppress the dissociation process also causes the breakdown of STIRAP. For a fixed motion along $r$, the passage time of 20 ps is the threshold to obtain the yield of almost 100%. The applicability of STIRAP to a system with continuum or $N$-level has been shown using a simple model.[49,50] On the other hand, the results shown in Fig. 13 are an example showing limit of STIRAP via continuum.[51]

The optimal electric field of laser pulses has been derived by using a monotonically convergent iteration algorithm developed by Zhu and Rabitz.[52] The algorithm converges monotonically and quadratically in terms of the neighboring field deviations, if the expectation value of the target operator $W$ is positive definite. In their original formulation, the electric field function and the dipole moment were not vectors. The convergence property is still valid even if we substitute the dipole interaction $-\boldsymbol{\mu} \cdot \boldsymbol{E}(t)$ for $-\mu E(t)$.

We now show the results of chirality transformation of $H_2POSD$ including the competing photodissocaition process. The initial state was set to $|0L\rangle$, and the target operator $W_R = |0R\rangle\langle 0R| + |1R\rangle\langle 1R| + |4R\rangle\langle 4R| + |6R\rangle\langle 6R|$, in which numbers indicate vibrational eigenstates, localizes the R-handed enantiomer of $H_2POSD$ around $\phi = 60$ degrees in the ground electronic state. We set the final time $t_f$ to 100 fs and two components of the electric field vector $E_X(t)$ and $E_Y(t)$.

Figures 14(a) and 14(b) show the $X$ and $Y$ components of the electric field obtained by the optimal control method, respectively. The electric field vector, which selectively controls the pre-oriented molecular chirality in the pure state picture, is elliptically polarized, and the direction of polarization changes with passage of time. The optimal electric field consists of two sets of pump–dump pulses. The first pump and dump pulses are applied from 5 to 40 fs, transferring populations between the $S_0$ and $S_1$ electronic states. The second pump and dump pulses operate from 70 to 100 fs, and the target state is created by the last dump process.

Figure 14(c) shows the time and frequency-resolved spectrum used to analyze the pump and dump processes and to clarify the time-dependent phase of the electric field. It can be seen that

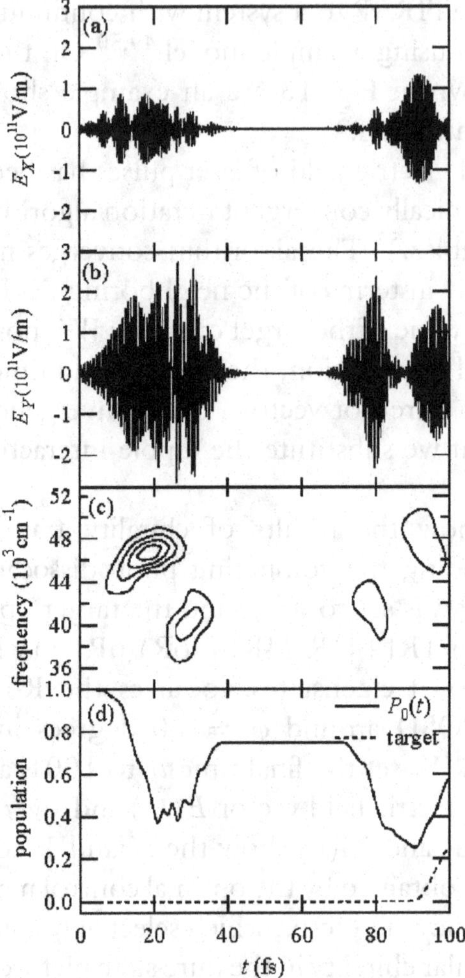


**Fig. 14.** Calculated optimal pulse and population changes for chirality transformation in the presence of the dissociation process. (a) and (b) denote the $X$ and $Y$ components of the optimal electric field, respectively. (c) Time and frequency-resolved spectrum of the electric field. The solid line in (d) denotes time propagation of the population in $S_0$, $P_0(t)$, and the dashed line denotes that of the expectation value of the target operator of the R-enantiomer.[48] Reproduced with permission from American Chemical Society.

the peaks of the four pulses are well separated from each other. This means that the optimal electric field vector consists of four separate pulses rather than two Raman pulses or an intra-pump–dump pulse. Furthermore, each pulse is chirped for timing of excitation or de-excitation of wave packet components.

Figure 14(d) shows the time-dependent population in $S_0$, and the expectation value of the target operator $W_R$. After the first pump–dump process, *ca.* 0.25 of probability is left in $S_1$, which dissociates. Similarly, after the second pump–dump process, *ca.* 0.17 of probability dissociates.

## 4. Conclusions

We have presented the results of theoretical studies on laser control of molecular chirality, which have been developed in our laboratory. First, we have focused on two fundamental aspects about laser control of molecular chiraity in a racemic mixture. The first aspect is that enantiomers should be pre-oriented for control of molecular chirality within the dipole approximation neglecting the magnetic dipole moment interaction. Once enantiomers are oriented, control of molecular chirality can be carried out by taking into account the photon polarization direction of the laser field since the direction of the dipole moment of a right-handed enantiomer is different from that of the corresponding left-handed one. Several schemes for control of molecular chirality are presented both in IR, visible or UV frequency ranges and ns to fs time ranges: an analytical treatment of preparation of enantiomers in the ground electronic state in an oriented racemic mixture, pump–dump control via an electronic excited state, control of molecular chirality in a randomly oriented racemic mixture using three polarization components of electric fields, and stimulated Raman adiabatic passage method. These control scenarios have been applied to $H_2POSH$, a simple real molecule having axial chirality. In this chapter, we have restricted our target molecules to those with axial chirality, and have omitted other classes of molecular chirality such as helical chirality that plays an important role in biochemistry.[53,54]

## Acknowledgments

The authors thank Professor J. Manz, Professor S. Koseki, Professor H. Kono, Professor Y. Ohtsuki, Professor L. González, Dr. H. Umeda and Dr. Y. Ohta, Dr. D Kroner, and Dr. F. Shibl for their critical comments and valuable suggestions. This work was partly supported by Grants-in-Aid for Scientific Research from the Ministry of Education, Science, Sports, Culture and Technology, Japan.

## References

1. B. M. Avalos, R. Babiano, P. Cintas, J. Jimenez, J. C. Palacios and L. D. Barron, *Chem. Rev.* **98**, 2391 (1998).
2. R. Noyori, *Asymmetric Catalysis in Organic Synthesis*, (Wiley, 1994).
3. D. J. Tannor and S. A. Rice, *Adv. Chem. Phys.* **70**, 441 (1988).
4. P. Gross, D. Neuhauser and H. Rabitz, *J. Chem. Phys.* **94**, 1158 (1991).
5. Y. Yan, R. E. Gillian, R. M. Whitnell, K. R . Wilson and S. Mukamel, *J. Phys. Chem.* **97**, 2320 (1993).
6. M. Sugawara and Y. Fujimura, *J. Chem. Phys.* **100**, 5646 (1994).
7. M. Sugawara and Y. Fujimura, *J. Chem. Phys.* **101**, 6586 (1994).
8. M. V. Korolkov, J. Manz and G. K. Paramonov, *J. Phys. Chem.* **101**, 13927 (1995).
9. M. Quack, *Femtosecond Chemistry*, Vol. 2, eds. J. Manz and L. Wöste (VCH. Weinheim, 1995) p. 781.
10. Y. Watanabe, H. Umeda, Y. Ohtsuki, H. Kono and Y. Fujimura, *Chem. Phys.* **217**, 317 (1997).
11. M. V. Korolkov, J. Manz and G. K. Paramonov, *Chem. Phys.* **217**, 341 (1997).
12. Y. Ohtsuki, H. Kono and Y. Fujimura: *J. Chem. Phys.* **109**, 9318 (1998).
13. B. Assion, T. Baumert, M. Bergt, T. Brixner, B. Kiefer, V. Seyfried, M. Strehle, and G. Gerber, *Science* **282**, 919 (1998).
14. Y. Teranishi and H. Nakamura, *Phys. Rev. Lett.* **81**, 2013 (1998).
15. Laser Control of Quantum Dynamics, Special Issue of *Chem. Phys.* Vol. **267** (2001).

16. D. Goswami and A. S. Sandhu, *Advances in Multi-photon Processes and Spectroscopy* Vol. 14 (World Scientific, Singapore, 2001) p. 132.
17. R. A. Grdon and Y. Fujimura, Coherent control of chemical reactions, *Encyclopedia of Physical Science and Technology* (Academic Press, San Diego, 2002) p. 207.
18. M. Shapiro and P. Brumer, *J. Chem. Phys.* **95**, 8658 (1991).
19. J. A. Cina and R. A. Harris, *J. Chem. Phys.* **100**, 2531 (1994); *Science* **267**, 832 (1995).
20. R. P. Duarte-Zamorano and V. Romero-Rochín, *J. Chem. Phys.* **114**, 9276 (2001).
21. A. Salam and W. J. Meath, *J. Chem. Phys.* **106**, 7865 (1997); *Chem. Phys.* **228**, 115 (1998).
22. J. Shao and P. Hänggi, *J. Chem. Phys.* **107**, 9935 (1997); *Phys. Rev. A* **56**, R4397 (1997).
23. M. Shapiro, E. Frishman and P. Brumer, *Phys. Rev. Lett.* **84**, 1669 (2000).
24. D. Gerbasi, M. Shapiro and P. Brumer, *J. Chem. Phys.* **115**, 5349 (2001).
25. E. Frishman, M. Shapiro, D. Gerbasi and P. Brumer, *J. Chem. Phys.* **119**, 7237 (2003).
26. P. Král, I. Thanopulos, M. Shapiro and D. Cohen, *Phys. Rev. Lett.* **90**, 033001 (2003).
27. Y. Fujimura, L. González, K. Hoki, J. Manz and Y. Ohtsuki, *Chem. Phys. Lett.* **306**, 1 (1999); *ibid* **310** 578, (1999).
28. L. González, K. Hoki, D. Kröner, A. Leal, J. Manz and Y. Ohtsuki, *J. Chem. Phys.* **113**, 11134 (2000).
29. Y. Fujimura, L. González, K. Hoki, D. Kröner, J. Manz and Y. Ohtsuki, *Angew. Chem. Int. Ed.* **39**, 4586 (2000); *Angew. Chem.* **112**, 4785 (2000).
30. Y. Fujimura, L. González, K. Hoki, J. Manz, Y. Ohtsuki and H. Umeda, Advances in *Multi-photon Processes and Spectroscopy* (World Scientific, Singapore, 2001).
31. K. Hoki, Y. Ohtsuki and Y. Fujimura, *J. Chem. Phys.* **114**, 1575 (2001).
32. K. Hoki, D. Kröner and J. Manz, *Chem. Phys.* **267**, 59 (2001).
33. K. Hoki and Y. Fujimura, in *ACS Books "Laser Control and Manipulation of Molecules"*, eds. A. D. Bandrauk, R. J. Gordon and Y. Fujimura (American Chemical Society, 2001) p. 32.
34. K. Hoki, L. González and Y. Fujimura, *J. Chem. Phys.* **116**, 2433 (2002).
35. A. Bkasov, T.-K. Ha and M. Quack, *J. Chem. Phys.* **109**, 7263(1998).
36. A. Bkasov and M. Quack, *Chem. Phys. Lett.* **303**, 547 (1999).
37. R. Berger, M. Gottselig, M. Quack and M. Willeke, *Angew. Chem. Int. Ed.* **40**, 4195 (2001).
38. P. Brumer, F. Frishman and M. Shapiro, *Phys. Rev. A* **65**, 015401 (2001).
39. Y. Inoue, *Chem. Rev.* **92**, 741 (1992).
40. D. P. Craig and T. Thirunamachandran, *Molecular Quantum Electrodynamics: An Introduction to Radiation-Molecules Interactions* (Academic press, 1984).
41. L. González, D. Kröner and I. R. Solá, *J. Chem. Phys.* **115**, 2519 (2001).

42. K. Hoki, L. González and Y. Fujimura, *J. Chem. Phys.* **116**, 8799 (2002).
43. K. Hoki and Y. Fujimura, *Chem. Phys.* **267**,187 (2001).
44. J. R. Kuklinski, U. Gaubets, F. T. Hiloe and K. Bergmann, *Phys. Rev. A* **40**, 6741 (1989).
45. U. Gaubetz, P. Rudecki, S. Schiemann and K. Bergmann, *J. Chem. Phys.* **92**, 5363 (1990).
46. K. Bergmann, T. Theuer and B. W. Shore, *Rev. Mod. Phys.* **70**, 1003 (1998).
47. Y. Ohta, K. Hoki and Y. Fujimura, *J. Chem. Phys.* **116**, 7509 (2002).
48. K. Hoki, L. González, M. F. Shibl and Y. Fujimura, *J. Phys. Chem. A* **108**, 6455 (2004).
49. C. E. Caroll, F. T. Hiroe, *Phys. Rev. Lett.* **68**, 3523 (1988).
50. V. S. Malinovsky and D. J. Tannor, *Phys. Rev. A* **56**, 4929 (1997).
51. T. Nakajima, J. Zhang and M. Elk, *Phys. Rev. A* **50**, R913 (1994).
52. W. Zhu and H. Rabitz, *J. Chem. Phys.* **109**, 385 (1998).
53. H. Umeda, M. Takagi, S. Yamada, S. Koseki and Y. Fujimura, *J. Am. Chem. Soc.* **124**, 9265 (2002).
54. K. Hoki, S. Koseki, T. Matsushita, R. Sahnoun and Y. Fujimura, *J. Photochem. Photobiol. A* **178**, 258 (2006).